SpringerBriefs in Earth System Sciences

Series Editors

Gerrit Lohmann, Universität Bremen, Bremen, Germany

Justus Notholt, Institute of Environmental Physics, University of Bremen, Bremen, Germany

Jorge Rabassa, Labaratorio de Geomorfología y Cuaternar, CADIC-CONICET, Ushuaia, Tierra del Fuego, Argentina

Vikram Unnithan, Department of Earth and Space Sciences, Jacobs University Bremen, Bremen, Germany

SpringerBriefs in Earth System Sciences present concise summaries of cutting-edge research and practical applications. The series focuses on interdisciplinary research linking the lithosphere, atmosphere, biosphere, cryosphere, and hydrosphere building the system earth. It publishes peer-reviewed monographs under the editorial supervision of an international advisory board with the aim to publish 8 to 12 weeks after acceptance. Featuring compact volumes of 50 to 125 pages (approx. 20,000—70,000 words), the series covers a range of content from professional to academic such as:

- A timely reports of state-of-the art analytical techniques
- bridges between new research results
- snapshots of hot and/or emerging topics
- literature reviews
- in-depth case studies

Briefs are published as part of Springer's eBook collection, with millions of users worldwide. In addition, Briefs are available for individual print and electronic purchase. Briefs are characterized by fast, global electronic dissemination, standard publishing contracts, easy-to-use manuscript preparation and formatting guidelines, and expedited production schedules.

Both solicited and unsolicited manuscripts are considered for publication in this series.

Öner Çetin
Editor

Agriculture and Water Management Under Climate Change

 Springer

Editor
Öner Çetin
Department of Agricultural Structures
and Irrigation
Faculty of Agriculture
Dicle University
Diyarbakir, Türkiye

ISSN 2191-589X   ISSN 2191-5903 (electronic)
SpringerBriefs in Earth System Sciences
ISBN 978-3-031-74306-1   ISBN 978-3-031-74307-8 (eBook)
https://doi.org/10.1007/978-3-031-74307-8

This Springer imprint is published by the registered company Springer Nature Switzerland AG
The registered company address is: Gewerbestrasse 11, 6330 Cham, Switzerland

If disposing of this product, please recycle the paper.

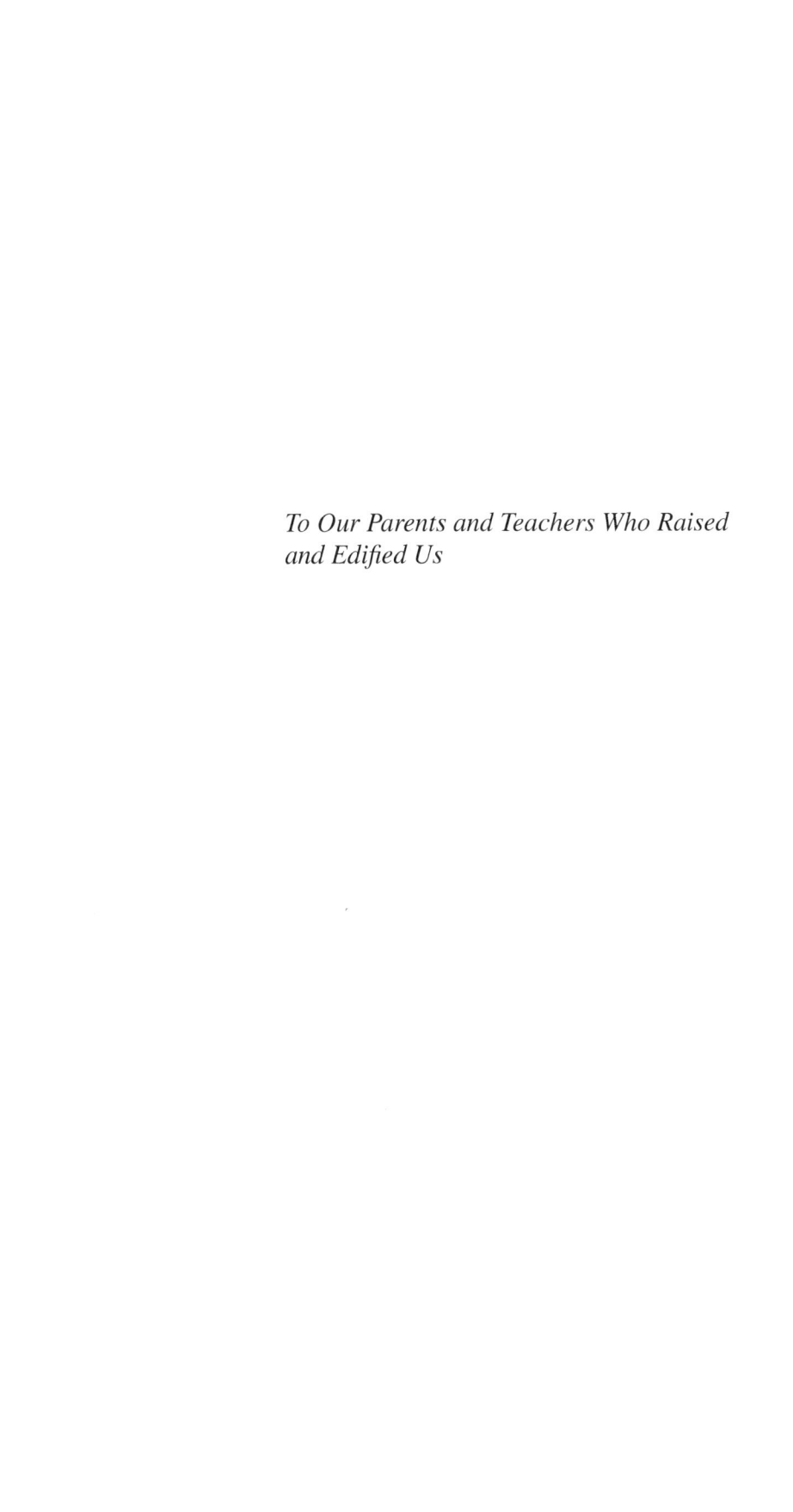

To Our Parents and Teachers Who Raised and Edified Us

Preface

Climate is one of the most important factors in ensuring the continuity and productivity of agricultural production depending on regional differences. Agriculture is a vital cycle that considers the entire ecosystem as a whole due to the provision of basic products required for the continuation of humanity and use of water and soil.

Despite the increase in food demand due to the world population, climate change stands out as the biggest threat to agriculture and food production. We are all experiencing the effects of climate change, which is on the agenda all over the world, with droughts or irregularities in the precipitation regime that cause excessive rainfall and floods one after another.

The impacts of climate change on agriculture can be listed as warmer and less rainy climatic conditions, increase in extreme meteorological events, decrease in water resources, increase in drought severity, deterioration of water and soil quality, degradation of ecosystems and decrease in biodiversity, shift in ecological areas, decrease in agricultural production and quality, increase in pests and diseases, fertilization and spraying problems and sustainable food security problems. In addition, climate change can affect forests, vegetation, land quality, clean water resources and biodiversity in many ways. Thus, climate change determines the direction and orientation of agricultural activities. Therefore, the risk posed by climate change on agricultural activities is quite high in terms of the unpredictability of its impacts.

All living things on Earth need food and shelter. Especially for human beings, food, water and energy are among the most essential needs. Both nutrition and food supply can be obtained from agriculture. Agriculture is directly dependent on climate, especially water and soil characteristics. In addition to feeding the growing population in a balanced way, it is also imperative to increase agricultural production in the face of food shortages and/or hunger. Agriculture includes both crop and animal production. Both of these agricultural production areas are directly dependent on climate and climatic characteristics. Today, climate change and global warming have brought these issues to the forefront, making it imperative that existing natural resources are used more effectively and efficiently.

Proper and conscious management of soil and water resources is essential for increasing agricultural production. It is globally recognized that human health, food security, welfare of societies and ecosystems are in great danger if strategically important soil and water resources are not managed effectively. Research and studies show that the lands to be opened for agricultural production in the world have reached the last limit, that it is not possible to open new agricultural areas, and that increasing income in agriculture can only be realized with a good utilization planning of existing lands.

On the other hand, the importance of information and communication technologies (ICT) and their application in agriculture in the context of the broader digital economy agenda should be promoted. Information and communication are essential for human, social and economic development. This is because ICTs offer farmers, including smallholders and family farming, the potential to access new technological content and sustainable and productive agricultural practices in markets with timely and accessible content.

Taking into account these reasons given above, efficient and sustainable use of natural resources, more economical use of diminishing water resources, adaptation to climate and climate change, some new approaches and use of technology in agriculture have become important in order to ensure the highest yield/production from agriculture. In this context, the chapters in this book include main topics such as climate crisis and its impact on crop production, resource utilization, use of technology in agricultural production, adoption and adaptation of technology in agricultural production, its convenience to agricultural production, difficulties and solution proposals and examples from different parts of the world, conservation agriculture, water and irrigation management, irrigation management and technology use in cotton. The different chapters in this book provide important information on the importance of past and current practices and the use of new technology, problems encountered and solutions to open a new horizon for those working in some important areas of agriculture, water and irrigation management. I believe that it will be useful for the user and those working in these fields.

Diyarbakir, Türkiye   Öner Çetin

Conflict of Interest The authors declared no potential conflicts of interest with respect to the research, authorship, and/or publication of all chapters.

Ethical Approval Not applicable for all chapters.

Contents

About the Editor

Öner Çetin is currently Professor at the Department of Agricultural Structures and Irrigation, Faculty of Agriculture, Dicle University, Diyarbakir City of Turkiye. After completing his Ph.D. at Çukurova University, Adana, Turkiye he has worked as a researcher on irrigation management in Soil and Water Research Institute under the Ministry of Agriculture from 1988 to 2006. He started as an academic staff at Dicle University, Diyarbakır, Turkiye, in 2006, and he continues at the same university.

He has participated to the courses on irrigation for four months in Utah State University, USA, and advanced courses on irrigation and soil management for two months in Israel and three months in Australia. Meanwhile, he has joined to some international conferences, symposiums, seminars and short training courses in many different countries (about 60). He has given also some international lectures. He is an active author having contributed more than 120 papers. In addition, he is an author for seven chapters and three books. He has completed four international conferences/workshops as a chairman. Lastly, he has actively taught irrigation scheduling, irrigation water management, soil-water relations, fertigation, and water productivity on training of farmers, agricultural technicians and engineers in national and international levels for long time. He is a member of the Irrigation Management Technical Committee of the Ministry of Agriculture and Forestry and a member of the Scientific Advisory Board of the Agricultural Strategy and Policy Development Center (TARPOL).

Abbreviations

AI	Artificial Intelligence
ATHOM	Flow Estimation and Optimization Model
BAU	Business As Usual
CA	Conservation Agriculture
CAP	Common Agricultural Policy
CEA	Controlled Environment Agriculture
CEMA	European Agricultural Machinery Association
CFS	Completely Fair Scheduling
CH_4	Methane
ÇKS	Farmer Registration System
CO_2	Carbon Dioxide
CPLM	Center Pivots and Linear Move
CRISPR	Clustered Regularly Interspaced Short Palindromic Repeats
CSA	Climate Smart Agriculture
CSB	Community Seed Banks
DEFRA	Ministry of Environment, Food and Rural Affairs in United Kingdom
DI	Deficit Irrigation
DMC	Domestic Consumption of Materials
DSI	General Directorate of State Hydraulic Works in Türkiye
DSS	Decision Support Systems
EC	European Commission
EPT	Environmental Protection Treatment
ERDF	European Regional Development Fund
ETa	Actual Evapotranspiration
ETc	Cop Evapotranspiration
EU	European Union
FAO	Food and Agriculture Organization of the United Nations
FDR	Frequency Domain Reflectometry
FFS	Farmer Field Schools
GCF	Green Climate Fund
GCOS	Global Climate Surveillance System

GDP	Gross Domestic Product
GFSI	Global Food Security Index
GHG	Green House Gases
GIS	Geographical Information System
GMM	Generalized Method of Moments
GMO	Genetically Modified Organisms
GPS	Global Positioning System
GRF	Global Environment Fund
GWh	Gigawatt Hour
IAASTD	International Assessment of Agricultural Knowledge, Science and Technology For Development
ICT	Information and Communication Technology
IFPRI	International Food Policy Research Institute
IMF	International Monetary Fund
INSTAT	Instituti I Statistikave (Institute of Statistics of Albania)
IoT	Internet of Things
IPC	Integrated Food Security Phase Classification
IPCC	Intergovernmental Panel on Climate Change
IPM	Integrated Pest Management
ISO	International Organization for Standardization
IT	Information Technology
ITA	International Trade Administration
IWP	Irrigation Water Productivity
Kc	Crop Coefficient
KOSGEB	Small and Medium Enterprises Development Organization in Türkiye
LEPA	Low Energy Precision Application
LWP	Leaf Water Potential
MAD	Management Allowed Deficit
MAS	Marker-Assisted Selection
MEGSİS	Spatial Real Estate System
MERNİS	Central Population Administration System
N_2O	Nitrous Oxide
NCT	Nature Conservation Tillage
NFIA	Netherlands Foreign Investment Agency
NGOs	Non-Governmental Organizations
NIS	National Institute of Statistics in Romania
NPV	Net Present Value
NRCS	Natural Resources Conservation Service
NREAP	National Renewable Energy Action Plan
O&M	Operation and Maintenance
OECD	Organisation for Economic Co-Operation and Development
OLS	Ordinary Least Squares
OWID	Our World in Data
PI	Production Intensity
PRD	Partial Root-Zone Drying

PSD	Protects Soil from Destruction
RS	Remote Sensing
RWC	Relative Water Content
SC	Stomatal Conductance
SCI	Site of Community Importance
SDG	Sustainable Development Goals
SSDI	Subsurface Drip Irrigation
SDI	Surface Drip İrrigation
SEM	Sustainable Ecosystem Management
SMC	Soil Moisture Content
SMS	Short Message Service
SUTEM	Irrigation Facilities Spatial Information System
SVM	Support Vector Machine
TAGEM	General Directorate of Agricultural Researches and Policies in Türkiye
TAKBİS	Land Registry And Cadastre Information System
TDR	Time Domain Reflectometry
TEV	Total Economic Value
TRGM	General Directorate of Agricultural Reform
TUBİTAK	Scientific and Technological Research Council of Türkiye
TUİK	Turkish Statistical Institute
UK	United Kingdom
UN	United Nations
UNEP	United Nations Environment Program
UNFCCC	United Nations Framework Convention on Climate Change
USA	United States of America
USD	United States Dollar
VRI	Variable Rate İrrigation
VRT	Variable Rate Technology
WDI	World Development Indicators
WEF	Word Economic Forum
WEP	Water Economical Productivity
WFP	World Food Program
WMO	World Meteorological Organization
WP	Water Productivity
WRI	World Resources Institute
WST	Water-Saving Technologies
WUA	Water User Association
WUO	Water User Organization
WWAP	World Water Assessment Program
WWTP	Wastewater Treatment Plant
WCM	Weed Control Measures

List of Figures

List of Tables

Chapter 1
Climate and Crop Production Crisis

Zoran Jovović⊙, Ana Velimirović⊙, and Neşe Yaman⊙

Abstract Climate change and its impact on the fate of humanity is one of the biggest challenges of today and a key topic of global discussion. Enormous economic activity, complex technologies, and rapidly growing ecologically destructive humanity have led to significant disruptions of the planetary ecological system. New scientific achievements and technological advancements have significantly benefited humanity, enabling easier exploitation of nature, and improving life quality. However, these advancements have also led to anthropogenic impacts on the environment, contributing to climate change through increased greenhouse gas emissions. Climate change adversely affects global agriculture and food security by causing extreme weather events, disrupting crop production, and elevating food prices. Agriculture both contributes to and suffers from climate change, as agricultural activities emit significant greenhouse gases and face challenges like soil infertility and water scarcity. Understanding climate change is crucial for planning adaptation strategies to ensure sustainable food systems. Technological innovations and sustainable practices are essential to enhance agricultural resilience. Future research should focus on sustainable practices, soil health, and climate smart technologies to mitigate climate change effects and ensure long-term food security.

Keywords Agriculture · Climate change · Climate resilience · Food security · Technological innovations

Z. Jovović (✉) · A. Velimirović
Biotechnical Faculty, University of Montenegro, Podgorica, Montenegro
e-mail: zoran.jovovic.btf@gmail.com

N. Yaman
Faculty of Agriculture, Department of Agricultural Structures and Irrigation, Dicle University, Diyarbakir, Türkiye
e-mail: nuzen@dicle.edu.tr

1.1   Introduction

New scientific achievements, advancements in technology, and production, espe-cially during the nineteenth and twentieth centuries, have brought many benefits to humanity. These advancements have helped humans to more easily shape nature, exploit its resources, and make life easier and better. Alongside all the positive changes, the rapid rise of human society has contributed to the emergence of a new factor—anthropogenic influence and its increasingly intense impact on both the living and non-living environment (Lewis & Maslin, 2015). This rapid development of human civilization, through excessive use of natural resources, has led to signifi-cant changes in the environment, seriously disrupted the natural balance of the planet, and altered the climate (Jovović & Kratovalieva, 2016). The increasingly extreme manifestations of climate are a consequence of anthropogenic actions on the atmo-sphere through increased greenhouse gas (GHG) emissions. As GHG levels in the atmosphere rise, many processes in nature induced by climate change are intensified.

Monitoring numerous anthropogenic impacts significant for specific contempo-rary climate systems has led to the understanding that the climate changes occurring in recent decades are predominantly driven by human influence (Zhang et al., 2007). Most temperature extremes recorded since the mid-twentieth century are likely the result of increased concentrations of anthropogenic GHG. Given that the global sea level and its temperature are constantly rising, and polar ice volumes are steadily decreasing, it can be reliably concluded that these phenomena are not merely routine natural processes (Jovović et al., 2020a).

The Fifth Assessment Report of the Intergovernmental Panel on Climate Change (IPCC, 2013, 2014a, 2014b) provides concrete evidence that over the last century, global temperature has increased by 0.8 °C and sea levels have risen by over 20 cm. Although this data might not seem alarming at first glance, the impact of global warming is clearly observed in many physical and biological systems on Earth, with a distinct imprint of human activities. There is, thus, ample evidence that climate change is occurring and that agriculture will deeply feel its impact.

Without urgent actions aimed at limiting GHG emissions, global temperatures could rise between 2.8 and 5.4 °C by the end of this century. However, a more likely scenario is that temperatures will increase in the range of 1.8–4.0, or 0.2 °C per decade. The highest temperature increases have been recorded in Southern Europe, and this trend is expected to continue. Most regional climate models predict that by the end of this century, air temperatures in Southeast Europe will rise from 2.4 to 2.8 °C, and in the Western Balkans from 1.7 to 4 °C, depending on the effectiveness of measures aimed at reducing GHG emissions. This large range of projections is due to the unreliability of climate models on one hand and the uncertainty of future emission levels on the other (Jovović et al., 2016).

On the other hand, a reduction in precipitation levels in the future will pose a serious problem for many regions worldwide, with more frequent heatwaves, sudden storms, and floods expected. This means that natural disasters will become more frequent, food and water will become less available, ecosystems will be destroyed or degraded, species extinction will accelerate, and human health will be threatened. The occurrence of extreme climate changes is daily worsening the lives of an increasing number of people and causing enormous material damage. According to research by the Swiss institute Swiss Re (2021), the total economic damage caused by climate change is increasing annually by 24%, and if these trends continue, the total global economic damage will reach 11–14% of the world's total economic output by 2050, or about 23 trillion USD.

To meet the food needs of the planet's growing population, it will be necessary to increase global food production by 60% by 2050, and by 100% in developing countries. However, this will not be an easy task (Jovović et al., 2016). The failure of climate action, extreme climate manifestations, and biodiversity loss will be the three biggest challenges in the coming years (WEF, 2024). Because of these challenges, over 200 million people could be displaced from their homes in the next 30 years. The highest numbers of climate migrants are expected in Latin America, North Africa, Sub-Saharan Africa, Eastern Europe, South, Central, and East Asia, and the Pacific.

The fight against climate change is one of the most important priorities of international policy. To assess the impacts of climate change on agricultural production and the degree of its vulnerability and to prepare appropriate adaptation measures, it takes at least 5–10 years. Knowledge gained from scientific diagnoses, as well as the experiences of some developed countries, is essential for establishing medium- and long-term agricultural development plans. Therefore, for more effective adaptation to climate change, new scientific and technological solutions tailored to local conditions will be needed. The development of technology and advances in the field of innovation over the last decade have permanently changed traditional agriculture, and today there are more and more initiatives promoting climate smart agriculture as one of the best tools for overcoming problems caused by climate change. Numerous climate smart agricultural practices (precision agriculture, agroecology, irrigation technologies, and crop protection against natural disasters, etc.) enable farmers to mitigate the sector's vulnerability to climate change, optimize the use of water, chemicals, and energy, and increase productivity. Given that agriculture is highly sensitive to climate change, it is necessary to take urgent adaptation measures, because if delayed, the consequences of climate change could reach dramatic proportions and irreversibly affect our future.

1.1.1   Understanding Climate Crisis

Until about half a century ago, the climate was a neutral topic for analysis and discussion, with little chance of being misused. However, this is no longer the case, as climate change is now perceived as one of the greatest challenges for humanity, the environment, and the global economy. Despite the peak in debates about climate change, there is still no clear consensus on their existence, and the direction they will take in the future remains an open question. While some continue to debate, others are rapidly preparing strategies to combat them. Both sides agree that recent climate change processes are a consequence of human activities affecting the atmosphere through increased greenhouse gas (GHG) emissions (Milinčić et al., 2015). As the consequences of climate change become more intense, they must be widely recognized and well-documented. Therefore, understanding climate change is crucial for the further advancement of civilization and the planning of complex adaptation processes.

Climate changes on Earth have been occurring for millions of years, but the causes of these changes are still not fully understood. During this period, the Earth's climate has constantly changed, but these changes occurred much more slowly. The reason why they are so relevant today is that this is the first time they are occurring as a result of human influence and are happening much faster than living systems can adapt to them (Gearheard et al., 2010). Most climate models indicate that the climate has been changing very intensively in recent decades and that the world is approaching a critical threshold where the pace of climate change will accelerate even more dramatically. Between 1880 and 2018, 19 of the warmest years occurred in the last 20 years, with 2016 being the warmest year since temperature measurements began (National Climate Assessment, 2018). Given this, the concern that the planet's climate balance could be significantly and permanently disrupted is entirely justified (Jovović et al., 2016).

The theory of climate change has been one of the most dominant scientific topics over the past 30 years. The degree of understanding future changes in the climate system will depend on the strength of the accumulated scientific evidence. The global scientific community is unanimous on this issue—climate change is our reality. For more than a century, scientists worldwide have been providing humanity with evidence of weather extremes and the role of human activities in these processes, with key points including:

- Enormous increases in atmospheric carbon dioxide and other greenhouse gases compared to the pre-industrial period.
- Greenhouse gases absorb heat when present in the atmosphere.
- A rise in average global temperatures over the past century by at least 0.85 °C and a sea level rise of 20 cm.
- Significant changes in the Earth's climate system (reduced snowfall in the northern hemisphere, retreat of Arctic sea ice, and glacier retreat on all continents).
- Significant changes in weather patterns and an increase in extreme events (North and South America, Europe, and northern and central Asia are becoming wetter, while central Africa, southern Africa, the Mediterranean, and southern Asia are becoming drier, with more frequent heatwaves and intense rainfall accompanied by major floods.

Overall, the world is far from where it needs to be, and the long-term effects of climate change on agriculture are becoming increasingly unfavorable, threatening global food security.

The Earth's temperature is determined by the balance between the absorbed and reflected energy from the Sun. About two-thirds of the Sun's energy passes through the Earth's atmosphere, warming its surface, while the remaining portion is reflected back into the atmosphere as long-wave infrared radiation. Most of this long-wave radiation is absorbed by atmospheric gases, with only a small fraction escaping into space. The greenhouse effect is a natural mechanism essential for maintaining the Earth's thermal regime; without it, the average surface temperature would be about 35 °C lower, or around −20 °C (IPCC, 2013). Atmospheric gases capable of absorbing and emitting long-wave radiation and infrared radiation are known as greenhouse gases. The most prevalent greenhouse gases in the atmosphere are water vapor, carbon dioxide, methane, nitrous oxide, and ozone. Additionally, chlorofluorocarbons, hydrofluorocarbons, tetrafluoromethane, and others are present in smaller concentrations. As the concentration of these gases increases, so does the greenhouse effect, thereby raising the Earth's temperature (IPCC, 2014a).

Developed countries have historically been the largest emitters of greenhouse gases, consistently so since the second half of the eighteenth century. Over the last 250 years, the concentration of carbon dioxide (CO_2) has increased by 31%, and by 18% since 1960, significantly disrupting the atmospheric energy balance. This greenhouse gas is considered the most significant driver of global warming, and its concentration in the atmosphere heavily depends on human activities. The rise in atmospheric CO_2 concentration is a consequence of burning fossil fuels and converting natural habitats into urban and agricultural areas. Since the beginning of the industrial revolution, global atmospheric CO_2 concentration has more than doubled, from 180 to 380–400 ppm. In 2007, China became the world's leading emitter of carbon dioxide in total quantities (IPCC, 2014b), although per capita emissions in the USA are four times higher (Figueres et al., 2018). By 2044, the world could emit another trillion tons of carbon dioxide into the atmosphere, equivalent to the emissions produced between 1750 and 2017 (Scheelbeek et al., 2018). Significant contributions to these projected greenhouse gas emissions will come from China, India, South Africa, and Brazil, whose rapid economic development is closely linked to energy production. Scientists believe these issues will become more pronounced in the future, as global temperature rises and the warming of the world's oceans will reduce their capacity to absorb carbon dioxide. Continued deforestation and the conversion of forest land into agricultural and urban areas will significantly reduce vegetation that could absorb emitted carbon dioxide.

During the same period, the concentration of methane (CH_4) in the atmosphere has also been steadily increasing. Of the total annual methane emissions, 60% are of anthropogenic origin, with this trend expected to continue in the coming years. The largest emitter of methane is agriculture (primarily livestock) at 24.2%, followed by the energy sector (oil and gas extraction, coal mines, etc.) at 23.3%, landfills and waste at 12.6%, and natural methane sources (marshes, wetlands, etc.) making up the remaining 40% (Global Methane Tracker, 2022). Compared to the pre-industrial period, the concentration of methane has nearly tripled (from 700 to 1867 ppb).

With the increase in greenhouse gas (GHG) concentrations, the average annual air temperature rises. Weather conditions are becoming more extreme, and the occurrence of weather-related disasters is more frequent. Due to global warming, air temperatures are continually increasing, and drought periods are more frequent and prolonged. Everywhere on the planet will get warmer, but not equally. It is estimated that the impacts of global warming are greater on land than on oceans. The largest temperature increase in Europe is expected in its southern regions. The highest air temperatures in that area were recorded in the first decade of the twenty-first century, which was also the warmest decade globally since 1880, the beginning of modern temperature measurements (Trenberth et al., 2007). During this period, according to the World Meteorological Organization, record temperatures were measured in 44% of the world's countries. Today, the average global temperature is around 15 °C, although historical data indicates it has significantly fluctuated, being much higher or lower at times. As there are regional differences in temperature increase, the effects of global warming are not uniform worldwide. Poor and developing countries, particularly those in the tropical and subtropical zones (southern Africa and South Asia), will be the most affected (IPCC, 2007a). In these regions, due to a lack of financial resources to adequately respond to the consequences of climate change, the greatest agricultural losses are expected (Lobell et al., 2008). The weakest effects are anticipated in developed countries with temperate climates, not only because of their more favorable geographic locations but also due to their significantly greater adaptive capacity. In Western European countries, agricultural production is expected to increase due to higher temperatures, increased CO_2 concentrations, and the application of modern technologies. Positive effects of warming are already noticeable in Scandinavian countries. In contrast, Mediterranean and Eastern European countries have recorded reduced yields in recent years (IPCC, 2007a). Agriculture in the Balkans, one of the fastest-warming regions on the planet, is already seriously threatened by floods, droughts, and summer heatwaves (Iglesias et al., 2007).

Climate change is also affecting precipitation patterns worldwide. The number of dry days is increasing, but there is also an increase in precipitation during the winter months, often with a potential for flooding. Changes in precipitation regimes will cause droughts in some countries and floods in others, while many will face both droughts and floods. Each degree of global temperature increase will result in a 5–10% change in precipitation patterns and a 3–10% increase in rainfall during the rainy season (Jovović et al., 2020b). Droughts will create increasingly severe problems for farmers worldwide, as about 80% of global crop production occurs in rain-fed conditions. Additionally, extreme rainfall and floods will significantly affect yields.

Due to its direct dependence on climate, agricultural production is one of the most vulnerable human activities. Rising temperatures, changes in precipitation patterns, and increased frequency of extreme weather events seriously impact the productivity of most major crops. This increases the vulnerability and sustainability of agriculture, with obvious effects on global food security. The greatest negative impacts are expected in tropical regions, where agriculture is the primary activity and main source of livelihood for over 60% of the population in sub-Saharan Africa and about 40–50% in Asia and the Pacific (IPCC, 2007b). Climate change will particularly affect populations in rural and underdeveloped areas, as their adaptive capacity is extremely low due to depopulation and economic backwardness.

On the other hand, the Paris Agreement on Climate Change (21st UNFCCC Conference, Paris 2015) agreed to keep global temperature increases below 2°C or less compared to pre-industrial levels. In order to avoid worst-case scenarios, it is necessary for global carbon emissions to reach peak by 2020, zero levels between 2060 and 2080, and negative values by 2100. To reach the 1.5 °C target, global carbon emissions, including those from agriculture, must be significantly reduced, and the zero level reached by 2050 at latest (IPCC, 2018). Even if greenhouse gas emissions from fossil fuels were completely eliminated, emissions from the global food system would make it impossible to limit warming to 1.5 °C, and the goal of 2 °C would be difficult to achieve. In order to fulfill the obligations of the Paris Agreement, a rapid reduction of greenhouse gas emissions, major changes in the way of food production, and a completely different relationship between man and the rest of the planet are necessary. Due to the accumulated amount of gases in the atmosphere, many of these changes are likely to last for decades, seriously threatening global food security (Clarke et al., 2016). In order to avoid greater environmental, economic, and social consequences, more ambitious global plans for reducing greenhouse gas emissions are needed, which would include a large number of developing countries in addition to industrialized countries.

1.1.2 Climate Change Effects on Global Agriculture and Food Security

Climate change negatively affects many sectors with its phenomena such as drought, floods, hurricanes, and heat waves. The food industry is one of these sectors. Specifically, climate-induced changes cause significant damage to crops, disrupt food supply chains, and generally result in elevated food prices and global food shortages. Additionally, rapidly changing weather conditions adversely affect the growth and productivity of crops. In this context, the primary and secondary effects of climate change on agriculture lead to food crises.

In many regions, increased evaporation due to changing precipitation patterns and warmer temperatures is reducing the availability of water for irrigation, potentially reducing crop yields (de Lima et al., 2023). Global warming is expected to weaken soil health and ecosystem chains such as pollination gradually and to negatively affect food productivity in many regions on land and in the ocean by reducing the biomass of marine organisms. Additionally, food crises resulting from climate change also have economic repercussions. During periods when food availability decreases, food prices are affected, and the cost of food imports increases, negatively affecting both consumers and businesses.

According to Oxfam Climate Change, Nutrition, and Fighting Hunger Report (Oxfam, 2013), the average price of basic nutrients will more than double in the next 20 years compared to 2010 prices, and more importantly, Oxfam shows climate change as the cause of more than half of this increase. In the Oxfam Report, it is emphasized that climate change means yield losses.

Food security is influenced by numerous interconnected factors, requiring a shift in the approach to food security in response to climate change. Research indicates that climate change significantly impacts food security, on both national and international agendas. Governments, regional, and international organizations are addressing the effects of climate change on human life in both short and long term. They have begun implementing measures to minimize or strategically adapt to its negative impacts. Key actors in this effort include the United Nations Framework Convention on Climate Change (UNFCCC), United Nations specialized agencies such as the Food and Agriculture Organization (FAO) and the World Food Program (WFP), as well as the joint initiative of the United Nations Environment Program (UNEP) and the World Meteorological Organization (WMO). IPCC, a scientific international organization established under the United Nations, is tasked with preparing the scientific basis and compiling studies on climate change. The UNFCCC facilitates political steps to address climate change, enabling coordinated global efforts to ensure food security in the face of a changing climate (IPCC, 2007b).

According to the 2015 data of Global Food Security Index, progress has been made toward ensuring food security in two-thirds of the countries evaluated. The average score of 109 countries increased by 1.2 points. While the region that made the most progress was the Middle East and North Africa Region (12 countries), a decline was observed in 85% of the 26 countries evaluated in Europe. The top three countries in the 2015 rankings are America, Singapore, and Austria, respectively. Turkiye ranked 39 in this index in 2014 and 2015. While the country with the most improvement is Egypt, the countries with the most decline include Sierra Leone, Israel, and Ukraine (Global Food Security Index, 2015).

It is estimated that if adaptation to climate change is not achieved for wheat, rice, and corn, it will negatively affect production. According to carried out a study, the increase in maximum temperatures, irregularities in rainfall distribution reduced wheat yields and had a significant negative impact on fertilization management (especially nitrogen) (Çetin et al., 2022). As a result, climate change is expected to significantly affect crop production in different regions. Projected impacts vary depending on product, region, and adaptation scenarios. While 10% of the predictions for the years 2030–49 predict an increase of more than 10% in product purchases compared to the twentieth century, 10% of the predictions indicate a product loss of more than 25% (IPCC, 2014c). Although the risks to agricultural products are expected to be more pronounced after 2050, this increase will depend on the level of increase in temperatures. In addition to increasing product demand, research indicates that the negative effects of climate change on food security are expected to outweigh any positive effects. Figure 1.1 presents estimates from these studies, illustrating the projected impacts on food security in light of changing climatic conditions and growing demand.

The effects of climate change on the four dimensions of food security on a global level have become an important reference source on the effects of climate change on food security by enriching these effects with examples periodically (FAO, 2008).

As widely recognized, the agriculture, fishing, and forestry are sectors sensitive to external conditions, and production processes are naturally exposed to the effects

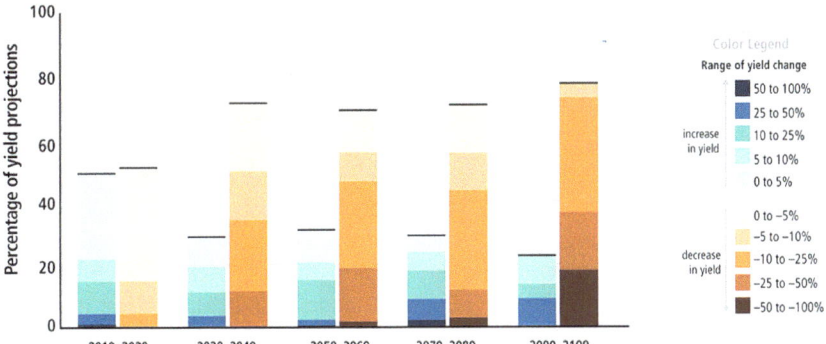

Fig. 1.1 Product yield estimates in IPCC Research (IPCC, 2014c)

of climate change. These effects are expected to be more favorable in temperate climates and negative in tropical climates (FAO, 2008). However, it is still unclear how the effects will be reflected at the regional scale.

According to the report, the impacts of changes in agricultural production and performance on food security are as follows: (i) impacts on food production will affect food supply at global and local levels. On a global scale, the increase in productivity in temperate climates will offset the decline in tropical climates. However, in many low-income countries that have limited financial means for trade and are dependent on their own production to meet their food needs, it will not be possible to offset any declines in domestic supply without increasing dependence on food aid; (ii) pressure on all forms of agricultural production is expected to affect livelihoods and access to food. The security and welfare of producer groups who are less likely to cope with the impacts of climate change, such as the rural poor in developing countries, are at risk.

Findings show that more frequent and more intense extreme weather events (droughts, heat and cold waves, severe storms, floods), rising sea levels, and fluctuations in etesian rain patterns affect not only food production but also food distribution. It directly affects infrastructure, the frequency of nutrition emergencies, and livelihoods in both rural and urban areas.

In addition, longer-term impacts are expected to occur as a result of changes in average temperature and precipitation values. According to research, climate change will affect agricultural production in the long term in the areas listed below: (i) Suitability of soil for different products, (ii) health and fertility of forests, (iii) distribution and productivity of seafood, (iv) vector and coincidence rates of different types of diseases and pests, (v) biodiversity and ecosystem functions of natural habitats, (vi) availability of quality water resources for plant, animal and seafood production and (vii) loss of cultivable land due to dryness (and associated salinity), losses of groundwater resources, and sea level rise.

Expected increases in average temperatures and precipitation regimes will not manifest themselves as steady and slow changes, but as increases in the frequency, duration, and intensity of heat waves and precipitation events (Gregory et al., 2005). Although areas that are sensitive to extreme weather events are known, there is no certainty about how expected changes in temperature and precipitation regimes will affect specific places (IPCC, 2014c). One of the most important effects of this is that planned investments in agriculture and climate-related livelihoods will be considered risky in the foreseeable future.

All aspects of food security use and price stability have a high potential to be affected by climate change. The increase in the distribution of marine fisheries catch potential toward upper latitudes indicates potential risks for food security by increasing the risks of decrease in supply, income, and employment in tropical countries (IPCC, 2014b).

Greenhouse gas emissions such as carbon dioxide, methane, and nitrous oxide resulting from the livestock sector affect climate change. According to FAO's 2015 data, emissions from the livestock sector correspond to 18% of global emissions (emissions equal to 7.1 billion tons of carbon dioxide) (FAO, 2017). 9% of global carbon dioxide emissions and 65% of nitrous oxide emissions are due to human activities and 35% of methane emissions originate from the livestock sector. Nitrous oxide emissions result from manure management, and methane emissions result from manure management and enteric fermentation.

An increase of 2 °C or more compared to the end of the century is expected to negatively affect production for many basic products (wheat, rice, and corn) in regions with tropical and temperate climates if climate change is not adapted. Projected impacts vary depending on product, region, and adaptation scenarios. While 10% of the predictions for the years 2030–49 predict an increase of more than 10% in production compared to the twentieth century, 10% of the predictions indicate a product loss of more than 25%.

Climate impacts in Europe are categorized into four main areas: temperature increases, water resource accessibility, floods, droughts, landslides, similar events, and sea level rise affecting coastal regions. Rising temperatures are disrupting plant and animal species distributions, especially in mountainous regions, altering animal behavior, and plant phenology. This shift threatens agricultural and livestock productivity while potentially increasing pests, invasive species, and human diseases (EEA, 2018).

Temperature rises exacerbate drought risks, particularly in Southern Europe, with potential desertification threats. Extreme heat events disproportionately affect agriculture, potentially reducing water availability in cities. Conversely, milder winters and fewer frost days are projected, affecting predictability and preparedness for climate events (EEA, 2018).

Access to water resources in Europe faces challenges, with decreases observed in Southern and Eastern Europe while other regions experience seasonal shifts. Summer droughts are expected to intensify, particularly impacting the Mediterranean and Central Europe, exacerbating fire risks and desertification (European Commission, 2014) (Fig. 1.2). Climate change is also projected to increase flood risks across

Fig. 1.2 Drought in the potato crop (30 days after emergence)

Europe, despite potential reductions in certain regions due to altered snow accumulation patterns. Heavy precipitation events are anticipated to rise, increasing flash floods and posing new challenges alongside existing flood risks (EEA, 2018).

Changes in climate variability will significantly affect agricultural productivity and production patterns. Longer growing seasons may benefit Northern Europe, enabling new crop suitability, while Southern Europe faces potential decreases in productivity due to extreme weather events and reduced water resources (Iglesias et al., 2010).

1.1.3 Contribution of the Agricultural Sector to Climate Change

The negative effects of climate change on agricultural activities are quite clear. Due to the water crisis, infertility of the soil, and extreme temperatures, the performance of agriculture becomes more difficult with each passing year. However, agricultural activities also have negative effects on the climate. Agricultural activities have a significant share of 20% among the factors that trigger the increase in greenhouse gases in the world (Pathak & Wassman, 2007). Activities such as energy consumption, production, fertilization, and pesticide use increase the release of carbon dioxide, methane, and nitrous oxide gases (Akalin, 2014). In addition, growing plants that require a lot of water in regions with low rainfall or using unconscious irrigation methods can be considered among the causes of the climate crisis, as they increase the water footprint.

Agricultural soils cause various emissions due to mineral fertilizer, animal manure, and also plant residues remaining in the soil. Balanced fertilization programs based on soil and plant analyses have a key role in reducing greenhouse gas emissions resulting from agricultural activities. Since mineral fertilizers and pesticides cause greenhouse gas emissions and pose a risk to health, reducing their use and providing the minerals the soil needs through composting are among the strategies that can be applied to both improve the soil and increase its carbon retention capacity. It should not be forgotten that emissions from mineral fertilizers occur not only after fertilization processes but also during the production, transportation, and application of fertilizer.

Soil cultivation, which has been used for different purposes since ancient times, is also among the causes of emissions. Agricultural methods that use mechanical tillage, such as plowing for seedbed preparation or plowing for weed control, can promote soil C loss through a variety of mechanisms and also disrupt the aggregate structure of the soil (Soares et al. 2005). It also stimulates short-term microbial activity with increased ventilation, resulting in high levels of CO_2 and other gases released into the atmosphere. In this way, plant residues mix with the soil, and decomposition conditions deep in the soil are generally more favorable than at the surface (Kladivko, 2001). In soil tillage systems that show the intensity of the effects on the soil, the highest CO_2 output occurred in conventional tillage systems where the soil was highly aerated, less gas output was detected in other reduced soil tillage systems, and the least CO_2 output occurred in no-tillage agricultural systems (Reicosky, 2003).

The greatest characteristic of organic agriculture is that it tends to improve the aliments and organic carbon required for the soil. Therefore, implementations include immediate recovery of animal manure to avert erosion of abundant topsoil, effective composting methods for crop remnants, and mixing crop waste with green manure. Progressing soil structure with these methods helps reduce greenhouse gas emissions. Research shows that 55–60% of the missing soil carbon can be recovered by progressing the soil structure (FAO, 2011a). Besides, reducing fossil fuel energy consumption in organic agriculture causes the soil to be less subjected to erosion and increases carbon attachment by ensuring the evaluating of aliments in the soil (FAO, 2011b).

In addition to being the lungs of the world, forests do much more than release oxygen. Working like a giant organ capable of photosynthesis, forests absorb carbon dioxide released into the atmosphere and convert it into oxygen. In this way, it reduces the destructive effects of greenhouse gases and ensures that the global temperature remains at ideal levels. It prevents drought by creating rain clouds and precipitation. Each forest creates a unique ecosystem and hosts various plants, fungi and algae, insects, animals, and microorganisms living above or below ground.

Deforestation contributes to approximately 12% of carbon dioxide emissions from human activities. This rate rises to 15% if you include tropical peats, which are now largely degraded and can contain up to ten times more carbon than forests (DeFries et al., 2016).

The livestock sector is responsible for 25–40% of anthropogenic methane emissions. This value appears consequently ruminant animals breaking cellulose throughout the enteric fermentation transaction. Roughly, 10% of methane making in livestock is produced from anaerobic manure saved. However, when animals graze, manure is dropped directly into the ground, reducing emissions from animal manure. Animal production affects global warming with its share of 9% of CO_2 emissions, 35–40% of CH_4 emissions, during the production, transfer, and 65% of N_2O emissions. Moreover, high temperatures and drought resulting from global warming are seen as threats to the survival of sustainable animal production systems and several ecosystems. Animal production is affected by pasture/forage plant number and quality, intensive feed raw material production and cost, disease, development and spread of pests, and changes in temperature and water availability.

While methane creates a primary greenhouse gas effect, the CO_2 effect is secondary. The heat capture capacity in the atmosphere of methane gas is 21 times greater than that of CO_2 gas. Compared to other gases its lifespan is shorter (Naqvi & Sejian, 2011).

. Ruminants are characterized by their structure having a special digestive system. In this way, they can easily digest low-quality cellulose-rich materials. As a result, ruminants are an important producer of methane. In fact, animals produce very small amounts of methane as an individual. For example, a cow produces approximately 80–110 kg of methane per year produced by ruminants. However, the large number they reach worldwide stands out rather than the amount of gas they produce. The effects of different sources on global methane emissions are given in Table 1.1.

One of the most effective ways to reduce emissions from agriculture is to meet the need for organic matter and minerals from outside, especially by composting vegetable residues near agricultural areas. Composting is the best method that can be used to improve the physical, chemical, and biological properties of the soil as well as to increase its carbon content. Biogas production and composting from organic waste are the most efficient strategies. In the last report, which included the greenhouse gas emissions of 28 countries of the European Union published by the European Environment Agency in 2018, it was stated that while the greenhouse gas emissions in 1990 were approximately 5650 million tons, the total emissions of these countries were reduced by taking the necessary precautions and relatively complying with the reduction and adaptation policies. Their value decreased by

Table 1.1 Different sources on global methane emissions (Naqvi & Sejian, 2011)

Natural resources (30%)	Human activities (70%)
Moist soil	Farming
Termites	Rice cultivation
Oceans	Coal mining
Gas hydrate	Biomass burning
	Landfills
	Animal manure
	Wastewater residues
	Natural gas and oil

24% in 2016, reaching approximately 4300 million tons. In the last 26 years of the research, Germany reduced its greenhouse gas emissions by 342.2 million tons, providing the most dramatic decrease in this area. This decrease was accompanied by the UK with 313.8 million tons. Countries with lower emissions, such as Lithuania, Latvia, Estonia, and Romania, have achieved great success by reducing their emissions by more than half. Austria, the Greek Cypriot Administration of Southern Cyprus, Ireland, Portugal, and Spain, on the other hand, failed to reduce greenhouse gas emissions and, on the contrary, increased them (EEA, 2018). Ways to reduce methane emissions in farm animals are as follows are listed (Naqvi & Sejian, 2011): (i) developing low methane-producing animals in genetic selection, (ii) improving animal nutrition with quality and strategic supplementation of essential nutrients and pasture management and use, (iii) providing appropriate care and health conditions for animals, (iv) reducing the roughage rate in their rations and increase the concentrate feed rate, (v) reducing methane production by making changes to ammonia and molasses in the diet, (vi) producing forage and pasture forage plants that provide less greenhouse gas emissions, (vii) using alternative forage plants and concentrated feeds with high tannin and saponin content, (viii) destroying protozoa in the rumen and microbial intervention in the rumen, (ix) reducing animal product production, (x) developing recombinant and immune technologies, (xi) reducing the number of animals by increasing the productivity of animals, (xii) using secondary plant components such as essential oils in animal nutrition, (xiii) adding vegetable oils to the diet, and (xiv) using probiotics that will suppress and compete with methanogenic microorganisms.

CO_2 is released into the atmosphere through events such as animal and human respiration and decay of organic matter. The energy needed in the food chain in CO_2 emissions constitutes 10% of the total emissions. It is stated that the most important source of CO_2 emissions is animal production and is equivalent to 9% of the total emissions. The source of the emissions is not the animal itself, but the CO_2 generated by the energy used in feed production, fertilizer processing, and transportation of products (Clark et al., 2001).

N_2O is the third most important greenhouse gas in terms of global warming, after carbon dioxide and methane. Although it makes up only 320 ppb of the Earth's atmosphere, it has a global warming potential approximately 300 times greater than that of CO_2. N_2O emissions are mainly of biogenic origin. N_2O is formed in soils and oceans worldwide as a result of nitrification and denitrification processes resulting from the N compounds ammonium and nitrate. These nitrogen compounds are released during the natural biogeochemical nitrogen cycle, but the most important and controllable are those released into the atmosphere by human activities. In fact, the amount of these compounds entering the biosphere has nearly doubled since the beginning of the fabrication age (Smith, 2010). The largest source is agriculture, where synthetic nitrogen fertilizers are currently used predominantly. Some N_2O also comes directly from combustion and two processes in the chemical industry: the production of nitric acid and the production of adipic acid, which is used in the production of nylon. Measures are being taken to curb industrial point source emissions of N_2O, but measures to limit or reduce agricultural emissions are inherently more difficult to manage, as the area under cultivation increases, the use of N fertilizer increases to feed the growing global society and to meet the current development of biofuels. Increasing N_2O emissions is unavoidable (Zaimoglu, 2019).

Approximately, 50% of corn and 80% of soybean produced in the world is used in animal production. Corn production requires the use of significant amounts of nitrogen fertilizer. In response to this necessity, nitrogen pollutes soil, water, and air. Additionally, significant amounts of diazotoxide gas are released from fertilizer tanks. Urine and manure accumulated as a result of grazing in the pasture or manure applied to the pasture are important N_2O sources. Globally, 5% of N_2O emissions come from plant production and 65% from animal production (Zaimoglu, 2019). It is recommended to reduce nitrous oxide emissions reducing the amount of nitrogen excreted through manure, reducing the number of animals and increase productivity, to reduce the amount of nitrogen given to the soil and reduce the amount of nitrogen in the ration.

Greenhouse gas emissions resulting from buffalo production activities constitute 9% of the total emissions. Greenhouse gas emissions from small ruminants constitute 6.5% of the total emissions. Generally, goats have a lower emission intensity than sheep in milk production because their milk yield is higher than sheep. Greenhouse gas emissions generated in pork production constitute 9% of the total emissions. This value is equivalent to approximately 668 million tons of CO_2. Approximately, half of the world's methane emissions from livestock manure come from pig manure. Emissions resulting from chicken production in the world are 606 million tons of CO_2, and this amount represents 8% of the emissions in the sector (Gorgulu et al., 2009).

Brazil, world leader in sugarcane agriculture, faces impending challenges from projected droughts in current cultivation regions, posing threats to both food production and bioethanol yields (Montgomery, 2010). Similarly, Korkmaz (2007) forecasts a potential 3–5 °C temperature decrease in Brazil by 2050, accompanied by an 11% increase in precipitation. While this may augment wheat (30%) and corn (16%) yields, soybean production could decline by 21% under these conditions.

Turkiye, situated in the Mediterranean Basin, anticipates adverse impacts on agricultural productivity due to climate change, predicting yield decreases ranging from 3.8% to 10.1% across all regions (Dellal et al., 2011). Cotton appears least affected, while corn faces significant vulnerability. Balcioglu et al. (2022) further illustrate these effects through their analysis of apricot production, noting decreased productivity and increased incidences of damaging events since 2000. Trends also indicate a notable temperature rise of 2.4 °C between 1980 and 2020, alongside a 20.2-day delay in apricot phenological phases. Turkoglu et al., (2016) reported that rising temperatures in Turkiye will shift plant phenology, advancing harvests in apple, cherry, and wheat, reducing grain quality and yield metrics, and increasing frost risk for fruit trees, impacting agriculture in developing nations.

The biological impacts of climate change on agricultural products were assessed under two scenarios. Agricultural products reliant on rainwater proved sensitive to both rainfall and temperature variations, whereas those under irrigation systems were predominantly influenced by temperature shifts (IFPRI, 2009). Due to climate change, productivity is projected to increase in middle and high latitudes, while declining significantly in tropical and subtropical regions. This shift may adversely affect the majority of rural populations and farmers. Additionally, inadequate nutrition due to climate-induced food shortages poses a significant health threat, particularly in vulnerable populations. The migration of people and animals from rural to urban areas is expected to reach 61% of the global population by 2025, causing environmental degradation, population growth, and food scarcity, leading to increased disease prevalence and mortality among migrating populations lacking adaptive capacity (Khasnis & Nettleman, 2005).

Apart from global climate change impacts, rapid population growth in densely populated nations like China and India is expected to escalate demand for agricultural products (Ozen, 2012). The rise in biofuel production has similarly driven up prices, particularly through higher grain demand. Some researchers argue that certain regions may benefit from climate change, emphasizing extended growing seasons and enhanced agricultural productivity facilitated by elevated CO_2 levels in Northern China, North America, and parts of Europe (Mendelsohn, 1999).

Agricultural production and food security in underdeveloped regions across Africa, Asia, and Latin America are affected by climate-related impacts. Given that more than one billion people worldwide suffer from malnutrition due to inadequate access to basic food sources (FAO, 2009), the profound implications of climate change on global agriculture are expected to intensify in the coming years.

On the other hand, wetlands, for example, store approximately 40% of global carbon; their preservation is crucial as drying releases stored carbon, significantly contributing to global warming (Erdem, 2013). Similarly, natural grasslands play a vital role in carbon absorption, but their conversion to agricultural lands without proper management results in substantial carbon emissions. Implementing policies to safeguard natural grasslands and making appropriate management changes can enhance carbon capture and sequestration capacities.

Deforestation, a key contributor to climate change, not only impacts global temperatures by altering energy absorption patterns but also reduces vital carbon stores found in forests. Forests annually sequester 3 to 5 tons of CO_2 per hectare while producing 8 to 13 tons of oxygen (Kiris & Toprak, 2009). Addressing climate change requires a long-term effort involving global cooperation, ranging from individual to national levels. While stopping climate change may not be feasible, mitigating its impacts remains an urgent global priority. Key strategies include: (i) sustainable planning of water resources for agriculture, (ii) development of efficient irrigation technologies and water management practices, (iii) promotion of drought-resistant seed varieties and sustainable land consolidation, (iv) advocacy for organic farming and adoption of good agricultural practices, (v) monitoring and mitigation of environmental pollution from agricultural activities, (vi) utilization of marginal quality waters for specific crops in agricultural irrigation, (vii) adoption of modern pressurized irrigation methods to reduce freshwater usage, (viii) implementation of volume-based water metering, (ix) provision of climate adaptation training from early ages, (x) breeding plants for stress resilience against factors like heat and drought, (xi) development and deployment of new crop patterns suited to changing climates, and (xii) reduction of emission sources through eco-friendly technologies.

1.1.4 Available Solutions for Adaptive Capacity of Agriculture and Climate Resilience

The present and future trajectory of agriculture is defined by the necessity for sustainable practices to address climate change, population growth, and environmental degradation. The prevailing discourse in agricultural development stresses modernization and increased productivity, which in turn poses environmental and social sustainability challenges and structural issues in many developing countries (Al-Agele et al., 2021; Losch, 2022). To achieve sustainability, technological advancements like micro-irrigation (Çetin & Kara, 2019), smart greenhouses, and agri-voltaic systems are crucial, along with the adoption of climate smart agriculture (CSA) and Conservation Agriculture (CA) practices (Al-Agele et al., 2021). CSA involves integrated approaches to managing farms for increased productivity, resilience, and reduced emissions. Practices such as precision farming, genetic modification for higher yields and disease resistance, and integrating renewable energy sources into farming are keys to optimizing food production while preserving the environment. The future of agriculture will require a shift toward sustainable land use, considering uncertainties in socio-economic drivers, climate change impacts, and the increasing demand for land for bioenergy production (Engstrom, 2016). The International Assessment of Agriculture, Science, and Technology for Development (IAASTD) highlights the importance of agricultural knowledge, science, and technology in addressing global challenges and ensuring equitable and sustainable development in the agricultural sector (Rivera-Ferre 2008). Various sustainable agricultural practices, technological innovations, policy support, community-based approaches, financial mechanisms, and research initiatives can support climate resilience in agriculture.

Key components of sustainable agricultural intensification involve genetic, ecological, and socio-economic aspects, focusing on improving yields, diversification, resource efficiency, and creating an enabling environment for sustainable practices. Sustainable agricultural practices encompass a range of eco-friendly techniques and approaches aimed at optimizing food production while preserving the environment and ensuring food security. These practices include precision farming, genetic modification for higher yields and disease resistance, crop rotation, organic farming, agroforestry, and integrating renewable energy sources into farming (Bhuyan et al., 2023). Novel approaches like climate smart agriculture, organic farming, and regenerative agriculture, along with practices such as integrated farming systems and precision agriculture, play a crucial role in promoting sustainable production and safeguarding agricultural sustainability (Muhie, 2022).

Conservation agriculture (CA) plays a crucial role in enhancing the adaptive capacity of agriculture and increasing climate resilience by promoting sustainable farming practices that mitigate land degradation, improve soil health, and enhance agroecosystem resilience to global change. CA principles, including minimum mechanical soil disturbance, permanent soil cover, and crop diversification, contribute to maintaining soil fertility, increasing farm yield, and reducing the need for external inputs, thus supporting sustainable food production in the face of climate challenges (Cárceles Rodríguez et al., 2022). Adaptive capacity of agriculture and increased climate resilience can be boosted by integrating trees with crops and livestock, providing multifunctional landscapes that promote biodiversity, nutrient cycling, and sustainability. Agroforestry systems help farmers adapt to extreme weather events, create resilient microclimates, combat climate change by sequestering carbon, and reduce greenhouse gas emissions, enhancing soil health and productivity (Wilson & Lovell, 2016). Sustainable resource utilization, improved soil health, and reduced environmental impact as key features of organic farming can increase climate resilience (Leifeld, 2012). Practices, such as crop rotations, organic fertilizers, symbiotic associations, and minimum tillage, soil fertility, and biodiversity are conserved; pollution is reduced, making it an environmentally friendly and sustainable agricultural approach. Organic farming also contributes to climate change adaptation by preserving environmental impact, improving food security, and addressing the challenges posed by climate change, especially in regions like Africa where its potential remains largely untapped (Wekeza et al., 2022). Additionally, organic strategies such as integrated nutrient and pest management ensure healthy products and reduced carbon concentrations in the atmosphere.

Technological innovations emerged as pivotal tools in modern agriculture enhancing agricultural efficiency, resilience, and sustainability in a changing climate. Precision agriculture harnesses cutting-edge technologies such as global positioning systems (GPS), geographic information systems (GIS), sensors, drones, remote sensing, data analysis, and machine learning enabling farmers to monitor and respond to within-field variability, leading to more efficient and precise farming techniques (Pedersen & Lind, 2017). Precision in fields and crop needs optimizes the application of inputs like water, fertilizers, and pesticides, maximizing yields and reducing costs and environmental impact by mitigating over-application of resources.

Resistant crop varieties, developed through biotechnology and traditional breeding techniques, are designed to withstand extreme conditions such as drought, heat, and salinity. These resilient crops offer farmers reliable yields in adverse environments, providing a crucial buffer against climate-induced production losses and ensuring food security in vulnerable regions (Velimirović et al., 2023). Climate smart irrigation systems represent another breakthrough, employing innovative techniques like drip and sprinkler systems. These systems efficiently deliver water directly to crops, minimizing wastage and optimizing moisture levels. Integrated with soil moisture sensors and weather forecasts, they adapt irrigation schedules in real-time, ensuring crops receive optimal hydration while conserving precious water resources (Driscoll et al., 2024).

Bridging the knowledge gap between scientific advancements and practical application on the farm is ensured through the role that extension services play. Strengthening agricultural extension services ensures that farmers receive timely information, training, and technical support on climate-resilient practices (Labarthe, 2009). Extension agents serve as facilitators of change, disseminating knowledge about new technologies, sustainable farming methods, and effective risk management strategies to enhance farmers' adaptive capacity. By leveraging local knowledge and fostering collaborative learning, community-based approaches empower farmers to adopt and disseminate effective practices. Farmer field schools (FFS) offer participatory training programs where farmers learn from each other and from experts about sustainable practices, pest management, and climate adaptation strategies. This peer-to-peer learning approach not only increases the adoption of resilient practices but also fosters a sense of community and shared responsibility among farmers. Community Seed Banks (CSBs) play a crucial role in preserving diverse crop varieties that are well-adapted to local conditions. Managed by local communities, these seed banks provide a vital resource for farmers, ensuring access to resilient seeds and maintaining genetic diversity (Jovović et al., 2020a; Velimirović et al., 2021).

Financial mechanisms and investments are critical for enhancing the adaptive capacity of agriculture and increasing climate resilience. Improving access to credit for smallholder farmers through microfinance institutions and rural banks offers affordable and accessible financial services, which empower farmers to implement climate-resilient measures and enhance productivity. Collaborations between governments, the private sector, and non-governmental organizations are vital for mobilizing resources and expertise in support of climate-resilient agriculture. These public–private partnerships drive innovation, scale successful initiatives, and provide farmers with the necessary support to adopt sustainable practices and technologies. International funding mechanisms like the Green Climate Fund (GCF) are crucial for supporting developing countries in their efforts to adapt to climate change. Accessing these funds helps finance large-scale adaptation projects and infrastructure development, ensuring that agricultural systems are better equipped to handle the impacts of a changing climate (Green Climate Fund, 2020).

Research and development are pivotal for advancing climate-resilient agricultural practices and ensuring long-term sustainability. Research to develop climate smart agricultural practices and technologies include breeding resilient crop varieties, enhancing soil health, and creating efficient water management techniques. Creating platforms for knowledge exchange among researchers, policymakers, and farmers for disseminating best practices and innovative solutions can be an efficient tool in creating climate-resilient system (Al-Kaisi & Lal 2020). Online portals, workshops, and conferences provide effective means for sharing valuable information and fostering collaboration across different stakeholders. Finally, monitoring and evaluation systems to assess the effectiveness of adaptation measures enable the refinement of strategies, learning from past experiences, and scaling up successful interventions, ensuring continuous improvement and resilience in agricultural practices.

1.1.5   Cutting-Edge Agricultural Technologies and Innovations

As climate change continues to pose significant challenges to global agriculture, cutting-edge technologies and innovations are becoming essential. Precision agriculture, drought-resistant crop varieties, and climate smart irrigation systems are transforming farming practices to enhance efficiency, reduce environmental impact, and increase resilience to climate variability (Chouhan et al., 2023). These advancements help mitigate the effects of climate change and also ensure sustainable and productive agricultural systems for the future.

GPS and GIS facilitate site-specific management practices, allowing for precise application of inputs such as fertilizers and pesticides, which optimizes resource use and minimizes environmental impact (Pedersen & Lind, 2017). Remote sensing technologies, including satellite imagery and drones, provide detailed data on crop health, soil conditions, and water availability. Drones equipped with multispectral cameras can quickly identify issues such as pest infestations or nutrient deficiencies, enabling timely interventions and reducing crop losses (Hunt Jr. & Daughtry 2018). Variable rate technology (VRT) allows for the precise application of inputs based on the specific needs of different areas within a field. VRT systems use data from soil sensors and yield monitors to adjust the rate of seeding, fertilization, and irrigation, improving efficiency and crop yields (Pedersen & Lind, 2017).

Biotechnology and genomics are revolutionizing agriculture by enhancing crop and livestock traits for better resilience and productivity. Genetically modified organisms (GMOs) are engineered for pest resistance, herbicide tolerance, and improved nutritional content. These modifications can lead to higher yields, reduced chemical use, and enhanced resilience to environmental stresses. Nutrition, CRISPR, and other gene-editing technologies enable precise modifications to accelerate the development of drought-tolerant, disease-resistant, and nutritionally enhanced varieties (Barman et al., 2020). Marker-assisted selection (MAS) uses molecular markers to identify desirable genetic traits in plants and animals (Boopathi & Boopathi, 2020).

Advanced irrigation systems are transforming agriculture by enhancing irrigation efficiency, water productivity, and crop quality. Drip irrigation delivers water directly to plant roots, minimizing waste and evaporation while ensuring a consistent water supply (Chouhan et al., 2023). Smart irrigation systems utilize sensors and weather data to optimize watering schedules, adjusting based on soil moisture, forecasts, and crop needs to conserve water. Aeroponics and hydroponics, which use nutrient-rich solutions in controlled environments, enable soil-less cultivation, resulting in faster growth, higher yields, and significantly reduced water use compared to traditional methods (Kumar et al., 2023).

Robotics and automation are revolutionizing agriculture by enhancing efficiency and precision. Autonomous tractors and agricultural machinery equipped with GPS, sensors, and machine learning algorithms can perform tasks such as planting, weeding, and harvesting with minimal human intervention, increasing efficiency, reducing labor costs, and allowing for precision farming (Pedersen & Lind, 2017).

Robotic harvesters equipped with vision systems and artificial intelligence can identify and pick ripe fruits and vegetables. These robots are particularly useful for labor-intensive crops and can work around the clock, increasing productivity and reducing post-harvest losses. Weeding robots use machine vision and artificial intelligence (AI) to distinguish between crops and weeds, precisely targeting and removing weeds, reducing the need for chemical herbicides, and minimizing crop damage (Li et al., 2022).

AI and data analytics are transforming modern agriculture by providing valuable insights and decision-making tools. Artificial intelligence and machine learning algorithms analyze vast amounts of data from various sources, including weather forecasts, soil sensors, and market trends, to provide predictive insights for farmers decisions on planning, harvesting, and marketing their crops (Hachimi et al., 2022). Decision support systems (DSS) integrate data from multiple sources to provide real-time recommendations for farm management. DSS can guide farmers on optimal planting dates, irrigation schedules, pest management, and nutrient application, improving overall farm efficiency (Chouhan et al., 2023). Farm management software platforms consolidate data from various farm operations into a single interface, providing tools for planning, monitoring, and analyzing farm activities, enabling farmers to optimize their resources and improve productivity.

Vertical farming involves growing crops in stacked layers or vertically inclined surfaces, often in controlled environments such as greenhouses or indoor facilities. This method maximizes space utilization, reduces water usage, and allows for year-round production regardless of external weather conditions. Controlled environment agriculture (CEA) systems use technology to create optimal growing conditions for crops. This includes controlling temperature, humidity, light, and CO_2 levels. CEA can significantly increase yields, reduce pest and disease pressures, and enable the cultivation of high-value crops in urban areas. Aquaponics combines aquaculture and hydroponics in a symbiotic system, whereas fish waste provides nutrients for the plants, while the plants help filter and clean the water for the fish (Kumar et al., 2023).

Blockchain technology provides a secure and transparent way to track and record transactions throughout the agricultural supply chain, enhancing traceability, reducing fraud, and ensuring the authenticity of products, from farm to fork. Digital marketplaces connect farmers directly with buyers, reducing the reliance on intermediaries and improving market access, offering real-time pricing information, streamlining transactions, and expanding market opportunities for smallholder farmers (Kraft & Kellner, 2022). The Internet of things (IoT) connects various farm devices and sensors, enabling real-time monitoring and control of agricultural operations. Smart farming systems can track soil moisture, weather conditions, crop health, and equipment performance, facilitating data-driven decision-making (Dagar et al., 2018).

1.1.6   Future Activities and Research Priorities

As agriculture faces increasing challenges from climate change and environmental degradation, advancing sustainable practices becomes imperative for long-term food security and ecosystem health. Advancing Sustainable Agricultural Practices, such as Integrated Pest Management (IPM) that combine biological, cultural, physical, and chemical tools to manage pests in an environmentally and economically sustainable manner are fundamental for creating climate-resilient agriculture (Fitt & Wilson, 2012). Enhancing soil health is critical for sustainable agriculture. Research should prioritize understanding soil microbiomes, developing practices that increase soil organic matter, and promoting regenerative agriculture techniques that restore and maintain soil fertility, such as cover cropping, crop rotation, and reduced tillage (Al-Kaisi & Lal 2020). Investigating the role of soil microorganisms in nutrient cycling, disease suppression, and plant health will promote development of microbiome-based soil management strategies. Expanding the adoption of climate smart agriculture practices is essential to increase productivity, enhance resilience, and reduce greenhouse gas emissions. This includes drought-tolerant crops, efficient irrigation systems, and agroforestry (Gupta et al., 2022).

Developing crops that can withstand extreme weather conditions is crucial for future food security. This requires genetic diversity exploration of underutilized and wild crop varieties to identify genes associated with stress tolerance, disease resistance, and nutritional improvements. Continued advancements in gene-editing techniques like CRISPR will create crops with enhanced resilience to drought, heat, salinity, and pests, furthermore, utilizing molecular markers to accelerate the breeding of climate-resilient crop varieties (Jovović et al., 2020b). Synthetic biology offers potential for creating new biological pathways and organisms that can address agricultural challenges. Therefore, research should explore the use of synthetic biology to develop bio-based fertilizers, biopesticides, and stress-resistant crops, as well as to enhance carbon sequestration in soils (Adams, 2016).

With increasing water scarcity, research should prioritize the development of water-efficient technologies and practices. Advanced irrigation systems, water recycling and harvesting techniques, and drought-resistant crop varieties are some of them (Evans & Sadler, 2008). Optimizing nutrient use efficiency elaborated through precision agriculture technologies tailor's nutrient applications to crop needs, soil testing and mapping, and the development of biofertilizers. Agricultural production can also benefit from merging sensors and AI to monitor and manage nutrient levels in real-time. Reducing energy consumption and increasing the use of renewable energy in agriculture using energy-efficient technologies, such as solar-powered irrigation systems, and the potential of bioenergy crops and anaerobic digestion of agricultural waste to produce biogas (Chouhan et al., 2023).

Big data and AI have the potential to revolutionize agriculture by enabling data-driven decision-making. Developing platforms that integrate diverse data sources, such as weather, soil, and crop data, and using AI algorithms to provide actionable insights for farmers are valuable tools for future agriculture (Hachimi et al., 2022).

Similarly, IoT can enhance farm management by connecting sensors, devices, and equipment to monitor and control agricultural processes in real-time (Dagar et al., 2018). Blockchain technology can enhance transparency and traceability in agricultural supply chains. Research should focus on developing blockchain applications that ensure the authenticity and safety of agricultural products, improve supply chain efficiency, and provide farmers with better market access and fair pricing (Kraft & Kellner, 2022).

Effective policy development and advocacy can foster sustainable agricultural practices and innovation. Advocating for policies that support research and development, provide incentives for adopting sustainable practices, and protect natural resources, research should analyze the impacts of existing policies and recommend improvements (Srivastav et al., 2021). Building the capacity of farmers, extension agents, and researchers is crucial for the adoption of new technologies and practices. Addressing global agricultural challenges requires international collaboration. Cross-border projects that share knowledge, technologies, and resources are essential for strengthening partnerships between governments, research institutions, and private sectors can facilitate innovation and the scaling of successful initiatives (Wagner & Linder, 2010).

1.2 Conclusions

Climate change is rapidly altering the environment, threatening global food security. There are serious indications that in the coming decades, food production worldwide will be significantly at risk. For these reasons, producing sufficient quantities of quality food will become one of the greatest global challenges. Although climate change is a global issue, its consequences are not evenly distributed. The biggest losers are likely to be underdeveloped and developing countries, which are already suffering significantly from the effects of climate change. Due to the global food system crisis, food prices will continue to rise, leading to increased hunger and malnutrition. This will primarily affect poor and vulnerable populations, increasing the number of people who suffer. Acute food shortages can lead to social unrest and conflict, as an increasing number of people fight for increasingly scarce resources. The lack of water and the continual loss of arable land will pose the most serious obstacles in addressing this pressing problem.

The key to successful adaptation lies in the hands of science and new technologies. It is science that must provide solutions to enable today's agriculture to adapt to its changing environment. To improve or eliminate the impacts of global warming on the agricultural system and minimize yield losses, the focus should be on technologies and agricultural practices that reduce the harmful effects of biotic and abiotic stresses and optimize plant growth. This will require the application of various conventional and unconventional practices, including adjusting sowing dates, managing nutrients and water, using plant hormones and osmoprotectants, as well as other techniques to mitigate the harmful impact of climatic factors. Developing new climate-resistant crop varieties will be one of the key factors for adaptation. Creating varieties that will be tolerant to drought, high temperatures, diseases, and other stressful situations will require not only new genetic sources but also knowledge from functional genomics and transgenic approaches, biotechnology, nanotechnology, and artificial intelligence. The resilience of the agricultural sector is crucial for food security, so greater investment in science and research can help farmers adapt more easily to climate change and ensure a sustainable future for agriculture.

References

Adams, B. L. (2016). The next generation of synthetic biology chassis: Moving synthetic biology from the laboratory to the field. *ACS Synthetic Biology., 5*(12), 1328–1330.

Akalin, M. (2014). The Effects of climate change on agriculture: Adaptation and mitigation strategies to eliminate these effects. *Hitit University Social Sciences Institute Journal., 7*(2), 351–377.

Al-Agele, H. A., Nackley, L., & Higgins, C. W. (2021). A pathway for sustainable agriculture. Sustainability, 13(8), 4328.

Al-Kaisi, M. M., & Lal, R. (2020). Aligning science and policy of regenerative agriculture. *Soil Science Society of America Journal, 84*(6), 1808–1820.

Balcioglu, Y. E. (2022). Effects of climate change on apricot yield and geographical distribution in Malatya province. *Journal of Environment and Nature Research, 4*(2), 119–146.

Barman, A., Deb, B., & Chakraborty, S. (2020). A glance at genome editing with CRISPR-Cas9 technology. *Current Genetics, 66*, 447–462.

Bhuyan, S., Laxman, T., Saikanth, D., & Badekhan, A. (2023). Advanced farming technologies for pollution reduction and increased crop productivity. In A Bhatia (Ed.), *Advanced farming technology*. (194–214)

Boopathi, N. M., & Boopathi, N. M. (2020). Marker-assisted selection (MAS). In N. M. Boopathi (Ed.), *Genetic Mapping and Marker Assisted Selection: Basics, Practice and Benefits* (pp. 343–388). Springer.

Cárceles Rodríguez, B., Durán-Zuazo, V. H., Soriano Rodríguez, M., García-Tejero, I. F., Gálvez Ruiz, B., & Cuadros Tavira, S. (2022). Conservation agriculture as a sustainable system for soil health: A review. *Soil Systems, 6*(4), 87.

Chouhan, S., Kumari, S., Kumar, R., & Chaudhary, P. L. (2023). Climate resilient water management for sustainable agriculture. *International Journal of Environment and Climate Change, 13*(7), 411–426.

Clark, H., de Klein, C., & Newton, P. (2001). Potential management practices and technologies to reduce nitrous oxide, methane and carbon dioxide emissions from New Zealand agriculture. *Ministry of Agriculture and Forestry, New Zealand*, 85.

Clarke, W. E., Higgins, E. E., Plieske, J., Wieseke, R., Sidebottom, C., Khedikar, Y., Batley, J., Edwards, D., Meng, J., Li, R., & Lawley, C. T. (2016). A high-density SNP genotyping array for Brassica napus and its ancestral diploid species based on optimized selection of single-locus markers in the allotetraploid genome. *Theoretical and Applied Genetics, 129*, 1887–1899.

Dagar, R., Som, S., & Khatri, S. K. (2018). Smart farming-IoT in agriculture. In *2018* International Conference on Inventive Research in Computing Applications (ICIRCA) (pp. 1052–1056).

de Lima, G. N., Zuñiga, R. A. A., Ogbanga, M. M., Leal Filho, W., Vidal, D. G., & Dinis, M. A. P. (2023). Impacts of climate change on agriculture and food security in Africa and Latin America and the Caribbean. In L. Filho (Ed.), *Climate Change and Health Hazards* (pp. 251–275). Springer Nature Switzerland.

DeFries, R., Sharma, S., & Dutta, T. (2016). A landscape approach to conservation and development in the Central Indian Highlands. *Regional Environmental Change, 16*(1–3). Springer Berlin Heidelberg.

Dellal, I., McCarl, B. A., & Butt, T. (2011). The economic assessment of climate change on Turkish agriculture. *Journal of Environmental Protection and Ecology, 12*(1), 376–385.

Driscoll, A. W., Conan, R. T., Marston, L. T., Choi, E., & Mueller, N. D. (2024). Greenhouse gas emissions from US irrigation pumping and implications for climate-smart irrigation policy. *Nature Communications, 15*(1), 675.

EEA. (2018). Annual European Union greenhouse gas inventory 1990–2016 and inventory report 2018. UNFCCC Secretariat. Retrieved from https://www.eea.europa.eu/publications/european-union-greenhouse-gas-inventory-2018 (Access date: 17 June 2024).

Engstrom, K. (2016). Pathways to future cropland: Assessing uncertainties in socio-economic processes by applying a global land-use model *(Doctoral thesis, Dept. of Physical Geography and Ecosystem Science*, BECC: Biodiversity and Ecosystem services in a Changing Climate).

Erdem, O. (2013). Importance, functions and values of wetlands. In *Wetlands Book* (p. 67). Ministry of forestry and water affairs, General Directorate of Nature Conservation and National Parks.

European Commission. (2014). Sixth national communication and first biennial report from the European Union under the UN Framework Convention on Climate Change (UNFCCC). Luxembourg: European Union.

Evans, R. G., & Sadler, E. J. (2008). Methods and technologies to improve efficiency of water use. *Water Resources Research, 44*(7).

FAO. (2008). *Climate change and food security: A framework document*. Food and Agriculture Organization of the United Nations.

FAO. (2009). Coping with a changing climate: Considerations for adaptation and mitigation in agriculture. *Environment and Natural Resources Management Series 15. Food and Agriculture Organization of the United Nations,* Rome

FAO. (2011a). Save and grow: A policymaker's guide to the sustainable intensification of smallholder crop production. Retrieved from http://www.fao.org/3/i2215e/i2215e.pdf (Access date: 10 June 2024).

FAO. (2011b). What is conservation agriculture? Retrieved from http://www.fao.org/ag/ca/1a.html (Access date: 3 June 2024).

FAO. (2017). Livestock Solutions for Climate Change. Retrieved from https://openknowledge.fao.org/server/api/core/bitstreams/0d178ab7-b755-4eb2-a6cd-05ba1db35819/content (Access date: 15 May 2024).

Figueres, C., et al. (2018). Emissions are still rising: Ramp up the cuts. *Nature, 564*, 27–30.

Fitt, G., & Wilson, L. (2012). Integrated pest management for sustainable agriculture. In D. P. Abrol & U. Shankar (Eds.), *Integrated pest management principles and practice*, 27–40.

Gearheard, S., Pocernich, M., Stewart, R., Sanguya, J., & Huntington, H. P. (2010). Linking Inuit knowledge and meteorological station observations to understand changing wind patterns at Clyde River, Nunavut. *Climatic Change, 100*, 267–294.

Global Food Security Index. (2015). Global food security index 2015. Retrieved from https://non ews.co/wp-content/uploads/2018/10/GFSI2015.pdf (Access date: 19 June 2024).

Global Methane Tracker. (2022). Reports: Methane and climate change. Retrieved from https://www.swissre.com/institute/research/topics-and-risk-dialogues/climate-and-natural-catastrophe (Access date: 10 June 2024).

Gorgulu, M., Koluman Darcan, N., & Göncü, S. (2009). Animal husbandry and global warming. In *5th National Animal Nutrition Congress*. Çorlu: Çukurova University.

Green Climate Fund. (2020). Green climate fund. Retrieved from https://www.greenclimate.fund/about (Access date: 20 June 2024).

Gregory, P. J., Ingram, J. S., & Brklacich, M. (2005). Climate change and food security. *Philosophical Transactions of the Royal Society, 2139–2148.*

Gupta, D., Gujre, N., Singha, S., & Mitra, S. (2022). Role of existing and emerging technologies in advancing climate-smart agriculture through modeling: A review. *Ecological Informatics, 71,* 101805.

Hachimi, C. E., Belaqziz, S., Khabba, S., Sebbar, B., Dhiba, D., & Chehbouni, A. (2022). Smart weather data management based on artificial intelligence and big data analytics for precision agriculture. *Agriculture, 13*(1), 95.

Hunt, E. R., Jr., & Daughtry, C. S. (2018). What good are unmanned aircraft systems for agricultural remote sensing and precision agriculture? *International Journal of Remote Sensing, 39*(15–16), 5345–5376.

IFPRI. (2009). Impact on agriculture and costs of adaptation. *Food Policy Report.*

Iglesias, A., Avis, K., Benzie, M., Fisher, P., Harley, M., Hodgson, N., Horrocks, L., Moneo, M., & Webb, J. (2007). Adaptation to climate change in the agricultural sector. *AEA Energy & Environment and Universidad De Politécnica De Madrid, 1,* 2020–2102.

Iglesias, A., Mougou, R., Moneo, M., & Quiroga, S. (2010). Towards adaptation of agriculture to climate change. *Regional Environmental Change, 159*–166.

IPCC. (2007a). Fourth Assessment Report of the Intergovernmental Panel on Climate Change (IPCC). Retrieved from https://www.ipcc.ch/assessment-report/ar4/ (Access date: 8 June 2024).

IPCC. (2007b). Intergovernmental Panel on Climate Change: Impacts, adaptation and vulnerability. Contribution of Working Group II to the Fourth Assessment Report of IPCC. Cambridge: Cambridge University Press.

IPCC. (2013). Climate change 2013: The physical science basis. Contribution of Working Group I to the Fifth Assessment Report of the Intergovernmental Panel on Climate Change. *Cambridge: Cambridge University Press.*

IPCC. (2014a). Climate change 2014: Impacts, adaptation, and vulnerability. Contribution of Working Group II to the Fifth Assessment Report of the Intergovernmental Panel on Climate Change. *Cambridge: Cambridge University Press.*

IPCC. (2014b). Climate change 2014: Mitigation of climate change. Contribution of Working Group III to the Fifth Assessment Report of the Intergovernmental Panel on Climate Change. *Cambridge: Cambridge University Press.*

IPCC. (2014c). Climate change 2014 synthesis report. *Intergovernmental Panel on Climate Change.* Retrieved from http://www.ipcc.ch/report/ar5/syr/ (Access date: 20 June 2024).

IPCC. (2018). Special report on global warming of 1.5 °C. Retrieved from https://www.ipcc.ch/sr15/ (Access date: 19 June 2024).

Jovović, Z., & Kratovalieva, S. (2016). Global strategies for sustainable use of agricultural genetic and indigenous traditional knowledge. In R. K. Salgotra & B. B. Gupta (Eds.), *Plant genetic resources and traditional knowledge for food security* (pp. 39–72). Springer.

Jovović, Z., Micev, B., & Velimirovic, A. (2016). Impact of climate change on potato production in Montenegro and options to mitigate the adverse effects. *Academic Journal of Environmental Science, 4*(3), 47–54.

Jovović, Z., Andjeloković, V., Pržulj, N., & Mandić, D. (2020a). The importance of preserving biodiversity for sustainable utilization of plant genetic resources. In N. Pržulj & V. Trkulja (Eds.), *From genetics and environment to food* (pp. 35–90). Academy of Sciences and Arts of the Republic of Srpska, Monograph XLI.

Jovović, Z., Andjeloković, V., Pržulj, N., & Mandić, D. (2020b). Untapped genetic diversity of wild relatives for crop improvement. In R. K. Salgotra & S. M. Zargar (Eds.), *Rediscovery of genetic and genomic resources for future food security* (pp. 25–65). Springer.

Çetin, Ö., Yildirim, M., Akinci, C., & Yarosh, A. (2022). Critical threshold temperatures and rainfall indeclining grain yield of durum wheat (Triticum durum Desf.) during crop development stages. *Romanian Agricultural Research, 39,* 247–257, DII 2067–5720 RAR 2022–48

Çetin, Ö., & Kara, A. (2019). Assessment of water productivity using different drip irrigation systems for cotton. *Agricultural Water Management* 223 (2019) 105693. https://doi.org/10.1016/j.agwat.2019.105693

Khasnis, A. A., & Nettleman, M. D. (2005). Global warming and infectious disease. *Archives of Medical Research, 36*(6), 689–696.

Kiris, R., & Toprak, S. (2009). The role of forests in climate change. In *TÜCAUM V. National Geography Symposium* (pp. 379–385).

Kladivko, E. J. (2001). Tillage systems and soil ecology. *Soil and Tillage Research, 61*(1–2), 61–76.

Korkmaz, K. (2007). Global warming and its effect on agricultural practices. *Alatarım Magazine, 6*(2), 43–49.

Kraft, S. K., & Kellner, F. (2022). Can blockchain be a basis to ensure transparency in an agricultural supply chain? *Sustainability, 14*(13), 8044.

Kumar, P., Sampath, B., Kumar, S., Babu, B. H., & Ahalya, N. (2023). Hydroponics, aeroponics, and aquaponics technologies in modern agricultural cultivation. In *Trends, paradigms, and advances in mechatronics engineering* (pp. 223–241). IGI Global.

Labarthe, P. (2009). Extension services and multifunctional agriculture. Lessons learnt from the French and Dutch contexts and approaches. *Journal of Environmental Management, 90*, S193–S202.

Leifeld, J. (2012). How sustainable is organic farming? *Agriculture, Ecosystems & Environment, 150*, 121–122.

Lewis, S. L., & Maslin, M. A. (2015). Defining the Anthropocene. *Nature, 519*, 171–180.

Li, Y., Guo, Z., Shuang, F., Zhang, M., & Li, X. (2022). Key technologies of machine vision for weeding robots: A review and benchmark. *Computers and Electronics in Agriculture, 196*, 106880.

Lobell, D. B., Burke, M. B., Tebaldi, C., Mastrandrea, M. D., Falcon, W. P., & Naylor, R. L. (2008). Prioritizing climate change adaptation needs for food security in 2030. *Science, 319*(5863), 607–610.

Losch, B. (2022). Decent employment and the future of agriculture: How dominant narratives prevent addressing structural issues. *Frontiers in Sustainable Food Systems, 6*, 862249.

Mendelsohn, R. (1999). Measuring the effect of climate change on developing country agriculture. *FAO Economic and Social Development Paper, 145*, 1–31.

Milinčić, M., Jovanović-Popović, D., Vujačić, D., & Peić, B. (2015). Impact of climate change on the global boundaries of sustainability. *Planning and normative protection of space and environment* (pp. 239–246). Association of Spatial Planners of Serbia, University of Belgrade - Faculty of Geography.

Montgomery, S. L. (2010). The forces driving global energy: *The 21st century and beyond* (1st ed.). Ankara, Turkiye.

Muhie, S. H. (2022). Novel approaches and practices to sustainable agriculture. *Journal of Agriculture and Food Research, 10*, 100446.

Naqvi, S. M. K., & Sejian, V. (2011). Global climate change: Role of livestock. *Asian Journal of Agricultural Sciences, 3*(1), 19–25.

National Climate Assessment. (2018). U.S. Global Change Research Program.

Oxfam. (2013). Un bouleversement croissant. Retrieved from http://www.oxfam.org/fr/rapports/un-bouleversement-croissant (Access date: 3 June 2024).

Ozen, N. E. (2012). Effects of financialization of product trade on agricultural product and food prices. *TEPAV Food and Agricultural Policy Research Institute*, 2.

Pathak, H., & Wassman, R. (2007). Introducing greenhouse gas mitigation as a development objective in rice-based agriculture: I. Generation of technical coefficients. *Agricultural Systems*, 807–825.

Pedersen, S. M., & Lind, K. M. (Eds.). (2017). *Precision agriculture: Technology and economic perspectives* (pp. 52–53). Springer International Publishing.

Reicosky, D. C. (2003). Tillage-induced CO_2 emissions and carbon sequestration: Effect of secondary tillage and compaction. In *Conservation agriculture, environment, farmers experiences, innovations, socio-economy, policy.* 291–300.

Rivera-Ferre, M. G. (2008). The future of agriculture: Agricultural knowledge for economically, socially and environmentally sustainable development. *EMBO Reports, 9*(11), 1061–1066.

Scheelbeek, P. F. D., Bird, F. A., Tuomisto, H. L., Green, R., Harris, F. B., Joy, E. J. M., Chalabi, Z., Allen, E., Haines, A., & Dangour, A. D. (2018). Effect of environmental changes on vegetable and legume yields and nutritional quality. *PNAS, 115*(26), 6804–6809.

Smith, K. (2010). *Nitrous oxide and climate change.* Earthscan Ltd.

Soares, M. L. G., Tognella-De-Rosa, M. M. P., Oliveira, V. F., Chaves, F. O., Silva, C. M. G., Portugal, A. M. M., Estrada, G. C. D., Barbosa, B., & Almeida, P. M. M. (2005). Environmental changes in South America in the last 10k years: Atlantic and Pacific controls and biogeophysical effects: Ecological impacts of climatic change and variability: Coastal environments—Mangroves and salt flats. *Report to the Inter-American Institute on Global Change (IAI),* 62.

Srivastav, A. L., Dhyani, R., Ranjan, M., Madhav, S., & Sillanpää, M. (2021). Climate-resilient strategies for sustainable management of water resources and agriculture. *Environmental Science and Pollution Research, 28*(31), 41576–41595.

Swiss Re Institute. (2021). Retrieved from https://www.swissre.com/media/news-releases/nr-20211214-sigma-full-year-2021-preliminary-natcat-loss-estimates.html (Access date: 18 June 2024).

Trenberth, K. E., Jones, P. D., Ambenje, P., Bojariu, R., Easterling, D., Klein Tank, A., Parker, D., Rahimzadeh, F., Renwick, J. A., Rusticucci, M., Soden, B., & Zhai, P. (2007). Observations: Surface and atmospheric climate change. In S. Solomon, D. Qin, M. Manning, Z. Chen, M. Marquis, K. B. Averyt, M. Tignor, & H. L. Miller (Eds.), *Climate change 2007: The physical science basis, contribution of Working Group I to the Fourth Assessment Report of the Intergovernmental Panel on Climate Change* (pp. 235–336). Cambridge University Press.

Turkoglu, N., Şensoy, S., & Aydin, O. (2016). Effects of climate change on the phenological periods of apple, cherry and wheat in Turkey. *International Journal of Human Sciences, 13*(1).

Velimirović, A., Jovović, Z., Perović, D., Lehnert, H., Mikić, S., Mandić, D., Pržulj, N., Mangini, G., & Finetti-Sialer, M. M. (2023). SNP diversity and genetic structure of "Rogosija", an old Western Balkan durum wheat collection. *Plants, 12*(5), 1157.

Velimirović, A., Jovović, Z., & Pržulj, N. (2021). From neolithic to late modern period: Brief history of wheat. *Genetika, 53*(1), 407–417.

Wagner, C., & Linder, R. (2010). The demand for EU cross-border care: An empirical analysis. *Journal of Management & Marketing in Healthcare, 3*(2), 176–187.

WEF (2024). Global Risks Report 2024. Retrieved from: https://www.weforum.org/publications/global-risks-report-2024/digest/ (Access date: 13 June 2024).

Wekeza, S. V., Sibanda, M., & Nhundu, K. (2022). Prospects for organic farming in coping with climate change and enhancing food security in Southern Africa: A systematic literature review. *Sustainability, 14*(20), 13489.

Wilson, M. H., & Lovell, S. T. (2016). Agroforestry-The next step in sustainable and resilient agriculture. *Sustainability, 8*(6), 574.

Zaimoglu, Z. (2019). Interaction of climate change and Turkish agriculture. *Climate Change Training Modules Series 7, Supporting Joint Efforts in the Field of Climate Change Project (iklimİN).* Retrieved from https://www.swissre.com/institute/research/topics-and-risk-dialogues/climate-and-natural-catastrophe (Access date: 15 June 2024).

Zhang, X., Zwiers, F. W., Hegerl, G. C., Lambert, F. H., Gillett, N. P., Solomon, S., Stott, P. A., & Nozawa, T. (2007). Detection of human influence on twentieth-century precipitation trends. *Nature, 448*(7152), 461–465.

Chapter 2
Adoption of Water-Saving Technologies in Agriculture

Andres Jaramillo Valencia

Abstract Water-saving technologies (WST) refer to tools, systems, and practices designed to reduce water consumption and enhance water efficiency in various applications, including agriculture, industry, and domestic use. These technologies aim to conserve water resources by minimizing waste, optimizing usage, and promoting sustainable water management. The adoption of water-saving technologies, particularly in agricultural advanced irrigation systems, presents equally significant opportunities and challenges. Precision irrigation systems such as sprinkler and drip irrigation can help conserve water and improve irrigation efficiency and water productivity. The adoption of water-saving technologies in agriculture is influenced by a complex interplay of factors that vary significantly between developed and developing countries. In many developing countries, traditional farming practices are deeply rooted, and there is often resistance to change. Adoption of modern technologies requires enormous efforts in capacity building and water policy drafting and implementation. Continued research, development, social work in farming communities, monitoring, and evaluation are key to ensure food security in a water-constrained future.

Keywords Precision irrigation · Sustainability · Technology · Water saving

2.1 Introduction

Water-saving technologies (WST) refer to tools, systems, and practices designed to reduce water consumption and enhance water efficiency in various applications, including agriculture, industry, and domestic use. These technologies aim to conserve water resources by minimizing waste, optimizing usage, and promoting sustainable water management. The adoption of water-saving technologies, particularly in agricultural advanced irrigation systems, presents equally significant opportunities and challenges. Several studies around the world have explored the factors influencing the adoption of these technologies, the benefits they offer, and the barriers that need

A. Jaramillo Valencia (✉)
Andres Jaramillo Consulting, 15 Watarrka Avenue, Fitzgibbon, QLD 4018, Australia
e-mail: andresja@hotmail.com

© The Author(s), under exclusive license to Springer Nature Switzerland AG 2024
Ö. Çetin (ed.), *Agriculture and Water Management Under Climate Change*,
SpringerBriefs in Earth System Sciences, https://doi.org/10.1007/978-3-031-74307-8_2

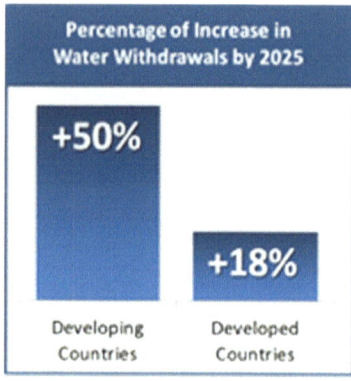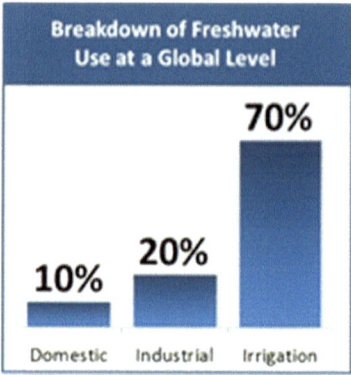

Fig. 2.1 Water usage statistics worldwide (UN, 2024)

to be overcome for widespread adoption; obviously, there are some unique characteristics and socio-politic environments around regions where those factors may vary.

On the other hand, water conservation in agriculture is vital for sustainable food production and environmental stewardship. As the pressure increases on freshwater resources, efficient water use helps to ensure long-term food security for a growing global population. Implementing water-saving techniques not only offers economic benefits to farmers through reduced costs and increased crop resilience but can also protect aquatic ecosystems. Water conservation practices are critical for adapting agriculture to the dramatic variations in climate particularly on precipitation patterns worldwide.

The agricultural sector faces increasing pressure to adopt water-saving technologies. Agriculture consumes approximately 70% of the Earth's water to irrigate cropland and produce food and other important crops (FAO, 2024). This review briefly discusses innovations used as water efficiency systems in farming around the world.

In this chapter, only agricultural technologies will be discussed, and water-saving technologies will be abbreviated as WST. Other non-agricultural examples include low-flow faucets, low-flow shower heads, water efficiency home appliances, advanced leak detection devices, water recycling, and reuse systems (Fig. 2.1)

2.2 Precision Irrigation Systems

There are irrigation methods that have been developed since the 1960s such as drip irrigation; this well-known system delivers water directly to plant roots, reducing evaporation and runoff. Drip Irrigation can be traced back to Ancient China when the use of buried clay pots filled with water (referred to as Ollas) was common as early as 100 BCE. It is an ancient technique practiced in various dry and arid parts of the world, especially in Africa, Pakistan, India, Iran, parts of Mexico, and Guatemala. In Pakistan and Afghanistan, it is known as "pitcher irrigation"; it consists of large clay pots with a wide bottom and narrow top buried in the ground and filled with water; the water is slowly released into the surrounding soil and absorbed by the roots of nearby plants, minimizing so the amount of water lost to evaporation (Fig. 2.2). Today, drip irrigation is widely considered one of the most valued innovations in the agricultural world since the 1930s, when the impact sprinkler was first introduced.

Agricultural sprinklers systems were first developed in New York, USA, in 1871 and, as mentioned above, perfected in the 1930s with the invention of the impact sprinkler head. Sprinklers can be defined as irrigation devices that distribute water through a network of pipes by spraying it into the air simulating natural rainfall. In time, other technologies in sprinkler systems like oscillating sprinklers, pop-up risers on timers, and agricultural-specific advancements like center pivot systems and low-pressure micro-sprinklers have given farmers other options in spray irrigation systems.

Center pivot irrigation was invented in 1940 in Colorado, USA. In the 1950s, the rolling pipe type of irrigation system was developed for large farms (Fig. 2.3). Micro-sprinklers, also known as micro-sprays, are small-scale irrigation devices that emit fine mist or spray of water, covering a smaller area than regular sprinklers with low pressure and volume. They provide targeted, low-volume water applications for permanent crops such as trees, orchards, and vineyards.

Subsurface drip irrigation (SDI) is another drip irrigation system whose characteristic is that it minimizes surface evaporation by applying water below ground. SDI has been a part of drip irrigation development since its beginning in the 1960s. Most early systems were primitive by current standards and consisted of holes or slits punched or cut into plastic pipe or discrete emitters punched into the pipe. The top five major advantages of SDI over on-the-ground installations are: (i) Permanent SDI requires less yearly labor than surface drip installation and increases the life expectancy of the system. (ii) Cropping with a dry soil surface has the potential to reduce the occurrence of soil-borne diseases and help control weed infestation. (iii)Dry soil in the furrows enhances trafficability and reduces soil compaction. (iv) Water and nutrients are used more efficiently, (v) Yields and certain quality components are often improved.(Anonymous, 2024b) (Figs. 2.2 and 2.3)

Moving irrigation systems, center pivots, and linear move (CPLM) are types of sprinkler systems that have unique features and are popular enough that they warrant a separate category of systems. Pivot irrigation delivers water in a circular area over a field. A lateral pipe is suspended across by structures named towers. Sprinkler heads are attached along the pipe or from the ends of hanging tubes. Pivot irrigation spans are commonly more than 300 m long (Fig. 2.4).

A linear move irrigation system is a mechanized method of delivering water to crops in a straight line. They are well-suited for rectangular or square-shaped fields. The system consists of a series of pipe spans supported by wheeled towers, like the center pivots, that move back and forth across a field. These spans have sprinklers on top or drop hoses and emitters to distribute water evenly across the crop.

Fig. 2.2 China Ancient drip irrigation drip irrigation system used in, Afghanistan, and Pakistan in the late seventeenth century (Anonymous, 2024a)

Fig. 2.3 Sprinkler irrigation mounted on a side-roll system (Anonymous, 2024c)

Although CPLM systems are used for almost any crop that does not grow taller than the overhanging lateral, they are typically used on larger acreage operations. Farmers commonly use CPLM systems for a large variety of crops including sugar cane, maize, potatoes, small grains, alfalfa, forages, and vegetables.

Fig. 2.4 Center pivots and linear move systems (CPLM) (ATLANTİS, 2024)

2.3 Other Water-Savings Technologies

In addition to the precision irrigation systems described above, there are several other water-saving technologies and practices in agriculture that can help conserve water and improve irrigation efficiency and water productivity. The level of adoption of these technologies tremendously varies by country, industry, farm size, farmer education level, and skills. The list below presents the most used complementary alternatives for water-saving technologies:

Soil moisture sensors: To monitor water content in soil to optimize irrigation scheduling.

Weather-based controllers: Used to adjust irrigation based on local weather conditions.

Remote sensing: Uses imagery to assess crop water needs across large areas.

Mulching: Applying organic or inorganic mulch to the soil surface helps reduce evaporation, maintain soil moisture, and improve water use efficiency.

Conservation tillage: Techniques like no-till or reduced tillage help maintain soil structure, reduce runoff, and increase water infiltration and retention in the soil.

Rainwater harvesting: Collecting and storing rainwater for irrigation can supplement water supplies, especially in regions with irregular rainfall patterns.

Drought-resistant crop varieties: Developing and planting crop varieties that are more tolerant to drought and require less water can help reduce overall water use in agriculture.

Terracing and leveling: Creating terraces on sloped land helps reduce runoff and erosion, allowing water to infiltrate the soil more effectively and reducing the need for additional irrigation. Flatter ground allows for more efficient irrigation and may save labor costs and improve crop yields.

Water reuse and recycling: Using treated wastewater or recycled water for irrigation can reduce the demand for freshwater resources.

Advanced crop management: Techniques such as deficit irrigation (DI), strategic under-watering at specific growth stages to improve water use efficiency, partial root-zone drying (PRD) (Zahoor & Mushtaq, 2023) and the use of drought-resistant varieties, both have positive results in water conservation when properly utilized.

Competition for water will force irrigation to operate under water scarcity in many parts of the world; DI and PRD, by reducing irrigation water use, can aid when water supply is drastically restricted. In field crops, a well-designed DI regime can optimize water productivity over areas when full irrigation will not be possible (Fereres & Soriano, 2007).

2.4 Reasonable Expectations from Water-Saving Technologies (WST)

There are two lines of arguments regarding the water-saving potential of WST. The first line of argument is that the adoption of WST results in net water savings and thereby eases the prevailing water-scarcity problems. The water savings are realized through substantial reduction in losses due to evaporation and inefficient field conveyance and distribution systems (Chand et al., 2020). This is one of the main reasons why governments embark on the promotion of WST. However, the farmers' rationale for adopting these technologies may be different from the government's policy objectives. Farmers may give more weigh to the other attributes of WST such as improvement in yield, reduction in labor requirement, improvement in output quality, modernizing of their farms, increased value of their farms and others, in their adoption decisions.

The second line of thought is that even though micro-irrigation technologies can result in water savings at the plot or field level (Çetin & Bilgel, 2002), it may not translate into net water savings at aggregation level such as the watershed or the basin (Molden et al., 2001). According to this line of thought, the net water savings could be only modest if the return flows (via deep percolation)—much of which goes to groundwater recharge—are considered useful.

Consequently, the adoption of WST may not automatically lead to water-saving at the basin level, unless enabling institutional and economic policy instruments are put in place that allow the equitable distribution or allocation of the *saved water*. So, when presented with an opportunity to reduce water consumption, while maintaining or even enhancing existing production, farmers see the water savings as a resource that can be reallocated by intensifying cultivation and extending the area under cultivation.

A study conducted in India on the impact that the National Mission on Micro-irrigation has had on water resources, reported that seven main impacts occurred. Out of these seven, the first reported impact was an increase in irrigation area from the same water source due to *water savings*. It found that the beneficiary farmers have brought an average of 8.4% of additional land under irrigation due to on-farm water saving from micro-irrigation.

Many farm-level studies of drip irrigation adoption focus on how to promote the further diffusion of drip irrigation through specific policies that will incentivize their adoption, while also enhancing economic and technical efficiency (Garb & Friedlander, 2014; Namara et al., 2007).

2.5 Factors Influencing Adoption of WST

The adoption of water-saving technologies in agriculture is influenced by a complex interplay of factors that vary significantly between developed and developing countries. In developed countries, the primary drivers include economic incentives, regulatory frameworks, technological advancements, and environmental awareness. Farmers in developed countries are often motivated by government subsidies, tax breaks, and grants that make the initial investment in advanced irrigation systems and other water-saving technologies more feasible.

In addition to these characteristics, stringent water use regulations and policies aimed at conserving natural resources encourage and subsidize the adoption of efficient water management practices. The availability of cutting-edge technology and access to skillful technical support also play crucial roles. Farmers in developed countries generally have better and easier access to education and training programs, which enhances their understanding and implementation of water conservation technologies. Environmental awareness and societal pressure to adopt sustainable practices further influence farmers' decisions.

In underdeveloped and developing countries, the factors influencing adoption are markedly different, often centering around economic constraints, lack of infrastructure, and limited access to information and technology. High upfront costs and limited financial resources are significant barriers, as many farmers cannot afford the initial investment required for advanced water-saving technologies. Inadequate infrastructure, such as unreliable electricity supply and poor irrigation networks, further hampers adoption.

The availability of credit and financing options is crucial but often lacking, making it difficult for farmers to invest in new technologies. Access to information is another critical factor; many farmers in developing countries are not aware of the benefits of water-saving technologies or how to implement them effectively, or simply, unable to afford the cost of adopting new technologies. Extension services, which play a vital role in educating and supporting farmers, are often under-resourced and unable to reach all potential users.

To illustrate this, lets analyze the cases of adoption of CPLM in developed and developing countries. In developed countries, the adoption of center pivot and linear move irrigation systems has been widespread due to their efficiency and effectiveness in water management. The advanced technology, availability of financial resources, and strong support infrastructure facilitate their implementation. Farmers in regions like the USA, Australia, and parts of Europe benefit from government subsidies, technical assistance, and research that enhance the efficiency and sustainability of these systems. The integration of modern technologies such as remote sensing, automation, and precision agriculture further boosts their adoption by allowing for better water management and reduced labor costs.

In contrast, the adoption of CPLM in developing countries faces significant challenges. High initial costs, farm sizes, level of education of farmers, and limited access to credit make it difficult for smallholder farmers to invest in these technologies.

Additionally, inadequate infrastructure, such as unreliable energy and water supplies and lack of technical support, hampers their widespread use. However, where implemented, these systems have shown promising results in improving water use efficiency and crop productivity. Nonetheless, the adoption of CPLM is mainly the privilege of wealthy farmers or corporate farming. Efforts by international organizations and governments to provide financial assistance, training programs, and infrastructure development are crucial in promoting these irrigation systems. Pilot projects and demonstration farms have played a vital role in showcasing these systems; their adoption is gradually increasing in regions such as Sub-Saharan Africa and Central Asia.

2.5.1 Socio-Cultural Factors

Socio-cultural factors also play a role in both contexts. In many developing countries, traditional farming practices are deeply rooted, and there is often resistance to change. Farmers may be skeptical of new technologies or lack the skills required to use them effectively; some farmers are used to using so much water that when they see the limited and controlled amounts of water applied by WST, they are reluctant to even give them a try. Peer influence and community leaders can either facilitate or hinder the adoption process. In contrast, farmers in developed countries usually face less resistance due to a greater openness to innovation and a higher level of technical proficiency.

Market conditions, including crop prices and the demand for water-intensive crops, also impact the adoption of water-saving technologies. In regions where water is scarce and expensive, farmers are more likely to adopt efficient irrigation methods to reduce costs and ensure the sustainability of their operations. On the contrary, in areas where water is cheap and abundant, there is less financial incentive to invest in water-saving technologies.

2.5.2 Environmental Factors and Institutional Support

Environmental factors, such as climate variability and the frequency of droughts, also influence adoption of WST. In regions prone to water scarcity, the need for efficient water management is more pressing, driving the adoption of technologies that can mitigate the impacts of limited water availability. On the contrary, in regions with abundant water resources, the urgency to adopt such technologies as a "water-saving" tool is much lower. Promoting WST in rain-fed agriculture, areas that are not serviced by canals or in hilly terrain has much better uptake than in regions where the idea is for farmers to switch from surface/gravity systems to modern pressurized WST.

Policy and institutional support are also critical elements across both developed and developing countries. Effective policies that promote water conservation, coupled with strong institutional frameworks and support, can create an enabling environment for the adoption of WST. In developing countries, international aid and development programs can play a significant role in assisting with the necessary resources and support to facilitate adoption. However, current subsidies, government procurement, and price restrictions lock farmers into low-value, undiversified farming systems and water-intensive crops. Poorly targeted subsidies in WST should be cut, while protecting the poorest, several donor entities subsidies, only benefit elite farmers or corporative farming.

Finally, there is a need for a careful analysis of the institutional setup of programs that are intended to support adaptation that goes beyond simply identifying barriers, but ones that explore how and why these barriers arise and continue to persist. It is important to unpack the network of actors and how their interests influence the evolution of government and non-government institutions. Traditional forms of institutional design may not be able to meet the requirements of those who are in most need of support and in a position to support the sustainable expansion of WST in underdeveloped countries (Jaramillo, 2024).

2.5.3 Economic Incentives and Subsidies

There are positives and negatives about economic incentives and subsidies toward the adoption of WST in agriculture. On the positive side, incentives and subsidies reduce the initial investment burden associated with purchasing WST and other water-efficient technologies. High initial costs are often mentioned as a barrier to adoption of WST in both developed and underdeveloped countries. By lowering the upfront costs, subsidies make it feasible for a broader range of farmers, including those with limited financial resources, to try these technologies.

Economic incentives in the way of grants, low-interest loans, or tax breaks, encourage investment in water-saving measures; measures like these can accelerate the adoption rate, leading to an increase in the installation of WST that can enhance water conservation and productivity. These economic incentives can lead to long-term benefits; users who adopt water-saving technologies often experience significant reductions in water usage, which depending on where they are can translate into lower operational costs and improved water resource management. This increased water productivity can result in higher crop yields and better-quality produce if the systems are designed, operated, and maintained properly, providing an economic boost to the agricultural sector.

Nevertheless, there are some negative aspects to consider about economic incentives and subsidies. One major concern is the potential for dependency-creation on subsidies. Farmers might become reliant on financial incentives, diminishing their motivation to seek out cost-effective solutions or innovate independently. This dependency can lead to inefficiencies and a lack of competitiveness. Also, the allocation of subsidies is not always efficient, sometimes leading to investments in technologies that are not optimal for specific regions or farming practices and sometimes not benefiting the most in need. Mismanagement, bureaucracy, and corruption can also divert funds away from their intended purposes, undermining the effectiveness of incentive programs.

The short-term focus of subsidies and financial incentives is another potential drawback. While subsidies can drive immediate action, that in itself does not necessarily mean adoption; they may not foster long-term commitment to the new practices, and therefore, the sustainability of the practice change is in question. Farmers might adopt water-saving technologies temporarily, for as long as they have subsidies or external support, without fully integrating them into their long-term farming strategies.

For instance, in a study in Maharashtra, India, where the promotion of WST is being pushed by the government, interestingly, some farmers have preferred unsubsidized, cheaper, lower-quality WST systems (drip) even when they were eligible for subsidies for more high-quality ones. They realized that the cost of growing tomatoes or onions per hectare with the unsubsidized lower-quality system was more economic than with the high-quality subsidized system.

Finally, inequity in access to subsidies also poses a challenge, as smaller or marginalized farmers may struggle to navigate complex application processes or meet eligibility criteria, leading to increased inequality within the agricultural sector.

While economic incentives and subsidies can be powerful tools for promoting the adoption of water-saving technologies in agriculture, their design and implementation must be carefully managed to avoid potential pitfalls. Well-targeted, transparent, and supportive programs can maximize the benefits, ensuring that these technologies are accessible, effectively used, and maintained for long-term sustainability.

2.5.4   Extension Services and Training Programs

Extension services are fundamental for disseminating knowledge about the most appropriate water-saving technologies for the community they are reaching out to. Extension services oversee and provide farmers with key information about the benefits, costs, drawbacks, and practical applications of WST. Through conducting demonstration projects that showcase the parts, installation, operation, and maintenance of the systems, as well as the effectiveness of WST, they validate the information presented in real-world, close to home, in the community's agricultural settings. These pilot demonstration projects help to build trust and confidence among farmers, as they can witness the benefits firsthand and can easier find an answer to their questions when presented with new technologies: *"what's in it for me?."*

Training programs equip farmers and installers with the necessary skills to assist with the installation or construction, operation, and maintenance of WST. Hands-on workshops and specific field training sessions ensure that farmers are competent in using these technologies effectively, leading to more sustainable adoption rates than otherwise. Structured training programs too play a crucial role in changing farmers' attitudes and behaviors toward water use. Farmers-to-farmers training is one of the most effective ways of transferring of technology. Peer learning and community building among farmers needs to be fostered in communities where WST are targeted to being introduced; when farmers share their experiences and success stories, it creates a supportive environment that fosters adoption of new practices.

The successful development, dissemination, and adoption of WST for small-holders depends on more than careful planning of research and the use of appropriate methodologies in extension. It largely depends on the timely formation of coalitions of key players, including key farmers as well as a range of key outsiders, researchers, and other stakeholders. Successful adoption of technology also depends on critical external factors—climatic events, market fluctuations, the availability of subsidies, and farm size (Cramb, 2000).

2.5.5   Farmer Perceptions and Attitudes

The adoption of WST is crucial for addressing water scarcity in the agricultural regions where it is already occurring. The way that farmers perceive practice change, although significatively varies by culture, does have common factors in both developed and developing countries. Farmer's technology perceptions and government regulations influence adoption behaviors.

The perceived ease of use of technology significantly contributes to farmers' adoption behavior regarding water-saving irrigation technologies (Wang et al., 2024). Farmers are more likely to adopt technology if they find it easy to use. The effect of the perceived usefulness of technology on farmers' adoption behavior is usually not significant; that is, farmers' perception of the usefulness of technology does not strongly influence their decision to adopt it.

Conventional research into farmer adoption of new technology explains the adoption-decision and its timing (early or late) primarily in terms of the decision maker's perceptions and inherent characteristics, with "innovators" at one extreme and "laggards" at the other (Rogers, 1995). Yet, farmer decision-making is generally more complex than that model. Farmers have multiple objectives including food security, adequate cashflow, income, children education, or purchasing assets or farming equipment. Many times, their adoption of new technology has nothing to do with their perception on the technology itself but in the disposable resources to uptake it and how that can affect the achieving of their other priorities.

Government regulations play a moderating role in the impact of the perceived ease of use of technology on the adoption of WST by farmers; government regulations can either facilitate or hinder the adoption of technology by farmers, depending on how they influence farmers' perception of the ease of use; the way that government deals with subsidies, training, information dissemination, regulation, certification, and standardization affects the way that farmers consider adoption of new technologies. Through the reduction of financial barriers, the provision of education and training, and by ensuring quality and reliability on WTS, adoption can be increased.

There are significant differences in the adoption behavior of WST between farmers with different farm sizes; large-scale farmers are influenced by government advocacy and technology subsidies, while smallholders are mainly influenced by the perceived usefulness of technology. So, factors influencing the adoption of new technology vary depending on the size of the farm.

Farmers are often persuaded that they can reduce the installation and operational cost of WST and ultimately crop production costs either through proper utilization of WST or via area expansion under these systems at their farms (which questions again the *water saved*). Similarly, policy formulators and development agencies should carefully consider crop zoning in devising their plans for the promotion of technology in the regions where WST are most needed. It is often recommended that the creation of "Farmers' Participation Studies" (FPS) involves local farmers in pilot studies to assess their willingness to adopt WST. In these groups, surveys and focus group discussions should be facilitated to understand farmers' perceptions, preferences, and concerns beyond the initial cost of installation of WST. FPS will help in designing appropriate extension strategies to promote sustainable adoption (Jaramillo, 2024).

The barriers for adoption of WST in the developing world do not differ much from country to country. For example, a study was conducted to find the constraints in adoption of drip irrigation for the farmers of Junagadh Taluka of Gujarat State, India (Pandya & Dwivedi, 2018), and these were the results obtained as influential for the adoption of the technology:

- High initial investment
- Problems in farming operations
- More maintenance required as compared to surface Irrigation
- Irrigation quantity by drip seems to be insufficient for crop growth
- Not satisfying after sales service
- Non-availability of technical knowledge and information about operating drip irrigation for different crops
- Problem of clogging of system due to water quality
- Non-availability of technical knowledge and information about operating drip irrigation for different crops
- Irrigation is to be done more frequently

In conclusion, the adoption of technology is influenced by many factors; some of the most impactful are the perceived ease of use, perceived usefulness, government regulation, and farm size.

2.5.6 Impacts of Adoption

As described in the previous sections, the adoption of WST in agriculture is a multi-faceted issue influenced by economic, technical, social, environmental, and institutional factors. As such, there are potential impacts in each of those areas depending on the scale of the adoption, the geographical setting, the community where its adoption takes place, and the support received from institutions and the government.

The adoption of WST in agriculture, such as drip irrigation and sprinkler systems, has been shown to improve agricultural production efficiency. For instance, drip irrigation, by delivering water directly to the root zone of plants, has a high potential to reduce water wastage and ensure that crops receive the optimal amount of water required for growth in a timely fashion. Tools such as this, also known as "precise irrigation," can enhance crop yields and improve the overall efficiency of agricultural production. For many, drip irrigation represents one of the most ideal technological solutions for today's agriculture. It is modern and can efficiently replace old and allegedly wasteful, gravity-based irrigation systems. It allows for more precise application of combined with fertilizers and therefore helps increase yields and productivity.

Nevertheless, there are few studies focusing on the real impact of drip irrigation. Drip irrigation is said to offer opportunities for farmers to increase their production and their income, help solve water-scarcity problems, and reduce waterlogging and to contribute to food security and poverty reduction. How exactly does drip irrigation achieve all these results is seldom explained (Venot et al. 2015). For example, the real impacts of drip irrigation kits, often promoted as a solution to poverty alleviation for smallholding farmers, are hardly ever evaluated over time. These kits have been enthusiastically promoted by donor organizations around the developing world, but their uptake has been vastly limited. Undoubtedly, more efforts into empirical research of the impacts of drip irrigation for small farmers, involving field measurements, are required. The five main impacts of adopting water saving technology in agriculture are given in Table 2.1.

2.5.7 Environmental Impacts

Assessing the extent of actual WST requires a comprehensive insight of the different elements of the water balance. Firstly, water that is "lost" from the cropped field through field runoff can be available to the nearest plot in the farm, and the water which is lost in deep percolation recharges the shallow aquifer, which can in turn be picked up by farmers using pumps and wells. Regarding evaporation from the soil profile, it depends on the crops whether they are distantly spaced such as fruit trees, fennel, or cotton castor, where evaporation can be significant; or field crops where soil evaporation evolves with cropping stages and the extent to which the crop canopy covers the soil. Likewise, evaporation is closely linked to weather conditions, and

Table 2.1 Main impacts caused by the adoption of water-saving technologies

Impact	Description
Increased water efficiency	Water-saving technologies reduce water wastage by delivering precise amounts of water directly to the plants, leading to more efficient water use and conservation
Enhanced crop yields and quality	Improved water management ensures that crops receive optimal water supply, which can lead to higher yields and better crop quality, enhancing agricultural productivity
Economic benefits	Reduced water consumption lowers water costs, and efficient use of fertilizers (e.g., through fertigation) can decrease input costs. Higher yields can increase farmers' income
Environmental sustainability	Reduced water usage and better management of resources minimize the environmental impact, conserving water bodies, reducing soil erosion, and lowering the risk of salinization
Resilience to climate change	Efficient water management helps farmers adapt to changing climate conditions, such as droughts and irregular rainfall patterns, ensuring more reliable crop production

it is lower in humid and sub-humid climates than it is in arid regions so, the extent of water savings through the use of water-saving techniques needs to be done on a case-by-case basis.

Drip irrigation has long been seen as a WST; in 1995, during the 5th International Micro-irrigation Congress (Lamm, 1995), the theme reflected environmental concerns: "Micro-irrigation for a changing world: Conserving resources/preserving the environment." Since then, its potential water savings have become an important feature for its promotion and irrigation modernization programs started to include the switching from surface irrigation to pressurized irrigation.

Linked to the use of WST is the understanding of the "rebound effect" (Berbel et al., 2015) which highlights the fact that farmers who save water per unit of land through the use of drip irrigation can, and often do, use the "saved" water to expand the area, hence resulting in no real water saved.

Several authors consider that adopting water-saving technologies promotes environmental sustainability by conserving water resources. They argue that these technologies help reduce the over-extraction of water bodies, including aquifers, preserving them for future use. Likewise, the argument for supporting WST as something positive for the environment is strengthened by the potential minimization of water wastage as runoff of agricultural chemicals into water bodies, decreasing the risk of water pollution and promoting healthier ecosystems (Lei & Yang, 2024).

While pressurized irrigation systems (drip, sprinkler, and CPLM) are recognized for their potential water-saving benefits and efficiency, they can also have some negative environmental impacts. Table 2.2 shows a list of five potential pros and cons of these water-saving technologies

Table 2.2 Pros and cons on the environment of the widespread adoption of water-saving technologies

Pros	Description	Cons	Description
Increased water efficiency	Reduces water wastage by delivering precise amounts of water, conserving water resources	**Soil salinization**	Continuous localized irrigation can lead to salt accumulation in the soil, affecting crop health
Enhanced crop yields and quality	Provides optimal water supply to crops, improving yields and quality, promoting sustainable farming	**Plastic waste**	Use of plastic components in these irrigation systems can lead to waste if not properly managed
Reduced water pollution	Minimizes runoff and leaching of fertilizers and pesticides, reducing water pollution and protecting ecosystems	**Water quality issues**	Improper management of fertigation can lead to leaching of chemicals into groundwater or surface water
Improved soil health	Techniques like conservation tillage and mulching improve soil structure and water retention	**Energy consumption**	Some water-saving technologies require significant energy input, increasing the carbon footprint
Climate change resilience	Helps farmers adapt to variable weather patterns, ensuring reliable crop production during droughts	**Ecological disruption**	Alteration of natural water flows can disrupt local ecosystems and hydrological patterns as well as aquifer level depletion in areas where large amount of groundwater is pumped

2.5.8 Social Impacts

Social relations and power dynamics in the farming context are decisive in determining the likelihood of a farming community to adopt WST. The underdeveloped world has millions of resource-poor farmers who are unable to bear their share of the cost of switching to WST.

Despite the technical transformation of WST to make them pro-poor, the well-to-do farmers still have significantly higher probability of adopting WST. In addition, farmers belonging to upper societal classes have more chance of adopting these technologies (Namara et al., 2007). In the social aspect side of things, the formation and development of agricultural cooperatives and collective economic organizations to help smallholders reduce costs and improve access to water-saving technologies should be promoted and supported.

There is also a question on innovation that should be asked: *How to provide effective and sustainable WST to smallholders?*. Catering for this sector of the farming population has not been in the top list of the priorities for WST manufacturers. Some level of government subsidies may still be required to make agriculture viable for these farmers though. Local support groups would be needed to take initiatives, develop some level of agro-entrepreneurship, and be consulted with for a type of WST that would work for them and who to train for its installation, operation, and maintenance.

2.5.9   Policy Considerations

To ensure the successful adoption of water-saving technologies in both developed and developing worlds, a comprehensive approach is essential. Firstly, financial mechanisms such as subsidies, low-interest loans, and grants should be established to alleviate the initial cost barriers for farmers and incentivize investments in sustainable practices.

Governments and international organizations must enhance awareness and knowledge through robust extension services, training programs, and demonstration projects to educate farmers on the benefits and operations of water-saving technologies.

Developing a reliable infrastructure, including efficient WST and maintenance services, is crucial for the effective implementation of these technologies. Policymakers need to design and enforce supportive regulatory frameworks that promote the adoption of water-saving technologies, streamline bureaucratic processes, and ensure coordination among various agencies. Bureaucratic hurdles, weak enforcement of existing water management regulations, and lack of coordination between various government agencies are fatal for the promotion of WST.

Addressing environmental and climatic challenges requires integrating adaptive technologies tailored to local conditions and promoting sustainable resource management practices.

2.6   Challenges and Barriers

The challenges faced specially by the developing and underdeveloped work for the adoption of water-saving technologies highlight the multifaceted nature of barriers to be overcome. Addressing these issues requires coordinated efforts involving financial support, educational initiatives, infrastructure development, policy reform, and consideration of environmental factors.

Simple adoption of WST will not result in water conservation because farmers are seen to expand the area under irrigation or shift to high-value crops (more water-intensive). A sad reality in the developing world is that the acceptance of innovations in agriculture, especially among smallholders, is not only challenging but also seldom

Table 2.3 Challenges and barriers for the adoption of water-saving technologies in the developing world

Challenge	Description	Examples of barriers
Financial constraints	High costs and limited access to financial resources impede investment in water-saving technologies	High initial costs, limited credit access, economic instability
Lack of awareness and knowledge	Insufficient information and training on water-saving technologies and their benefits	Limited dissemination, inadequate training, cultural resistance
Inadequate infrastructure	Poor existing infrastructure and lack of support services hinder technology implementation	Poor irrigation systems, unreliable water supply, inadequate maintenance and support
Policy and regulatory barriers	Weak policies and regulatory frameworks fail to incentivize or support technology adoption	Insufficient policies, bureaucratic hurdles, weak enforcement
Environmental and climatic challenges	Environmental variability and degradation reduce the effectiveness and adaptability of water-saving technologies	Rainfall variability, soil and water quality degradation, limited adaptability

rapid, and it is often unsuccessful (Van Rijn et al., 2012). Table 2.3 below summarizes the most common challenges and barriers for the adoption of WST

2.7 Conclusions

Water-saving technologies in agriculture can offer some solutions to the global water crisis. While initial implementation costs can be high, these innovations can provide long-term benefits in water conservation, crop yield, and environmental sustainability. Adoption of new technologies requires enormous efforts in capacity building and water policy drafting and implementation. Continued research, development, social work in farming communities, monitoring, and evaluation are key to ensure food security in a water-constrained future.

For over twenty years, low-cost low WST kits have been promoted by numerous organizations to smallholder farmers in developing countries. In many cases, these organizations have perceived these kits as potential game changers, for example, as pathways out of poverty on a mass scale. In the literature available, it is reported that such kits can lead to huge reductions in rural poverty (Venot et al., 2017). There are several articles such as the "Smallholder Irrigation as Poverty Alleviation Tool in Sub-Sharan Africa," by Woltering et al. (2011) and Burney and Naylor (2012), that argue that those kits can lead to huge reductions in rural poverty, based on studies undertaken on small pilot plots. However, as demonstrated by the independent studies reviewed here, there is no evidence of widespread adoption of drip irrigation kits among African smallholders, and the available evidence suggests their use is rarely, if ever, profitable and sustained over time.

In different WST projects in the developing world, the mechanism of entrusting post-installation service delivery to the private sector has not worked satisfactorily. This can be one of the main reasons whereby suboptimal performances of many of the installed water-saving technologies in agriculture. The lack of effectiveness of these post-installation arrangements hinders the momentum of the uptake of WST uptake as a single non-functional site creates more adverse effects as compared to several successfully operating units. Effective use of WST requires different farming practices such as land and seedbed preparation, planting, transplanting, irrigation scheduling, fertigation programming, weed control, insect and pest management, cultivation, up to harvesting crop, be likewise intervened and supported; only tackling the water efficiency aspects of cropping is not sufficient to achieve higher yields and therefore improve farmers' incomes.

In the literature reviewed for this chapter and for other technical documents on the adoption of WST for smallholders, it is enormously difficult to find proper independent impact evaluations of WST programs or projects financed by governments or donor institutions on their investments and sustainable adoption.

References

Anonymous. (2024a). Permaculture plants: Clay pot olla irrigation. Retrieved from: https://permac ultureplants.com/olla-irrigation (Access date: 08 June 2024).

Anonymous. (2024b) Subsurface drip irrigation (SDI) in the great plains. Western Kansas Research-Extension Centres. Retrieved from: www.wkrec.org/programs/irrigation_engineering/subsur face_drip/ (Access date: 30 June 2024).

Anonymous. (2024c) Wade rain irrigation systems. Retrieved from: https://www.waderain.com/ (Access date: 30 June 2024).

ATLANTIS. (2024) Atlantis center pivot and linear systems. Retrieved from: https://www.atlantis. com.tr/ (Access date: 02 July 2024).

Berbel, J., Gutiérrez-Martín, C., Rodríguez-Díaz, J. A., Camacho, E., & Montesinos, P. (2015). Literature review on rebound effect of water saving measures and analysis of a Spanish case study. Water Resources Management, 29, 663–678.

Burney, J. A., & Naylor, R. L. (2012). Small holder Irrigation as a Poverty Alleviation Tool in Sub-Saharan Africa. World Development, 40(1), 110–123.

Chand, S., Prabhat, K., Kumar, S., & Srivastava, S.K. (2020). Potential, adoption, and impact of micro irrigation in indian agriculture in India. Policy Paper No. 36. ICAR- National Institute of Agricultural Economics and Policy Research.

Cramb, R. (2000). Processes influencing the successful adoption of new technologies by smallholders. Working with Farmers: The Key to Adoption of Forage Technologies (11–22).

Çetin, Ö., & Bilgel, L. (2002). Effects of Different Irrigation Methods on Shedding and Yield of Cotton. Agricultural Water Management, 54(1), 1–15.

FAO. (2024). Water and one health. Retrieved from https://www.fao.org (Access date: 21 May 2024).

Fereres, E., & Soriano, M. A. (2007). Deficit irrigation for reducing agricultural water use. Journal of Experimental Botany, 58(2), 147–159. https://doi.org/10.1093/jxb/erl165

Garb, Y., & Friedlander, L. R. (2014). From transfer to translation: Using systemic understandings of technology to understand drip irrigation uptake. *Agricultural Systems, 128*, 13–24. https://doi.org/10.1016/j.agsy.2014.04.003

Jaramillo, A. (2024). Policy notes on recommendations for sustainable adoption of high efficiency irrigation systems in Punjab and Khyber Pakhtunkhwa Regions, *ADB Project Number*: (55225–001).

Lamm, F.R. (1995). Microirrigation for a changing world : Conserving resources/preserving the environment. Proceedings of the fifth International Microirrigation Congress American Society of Agricultural Engineers April 2–6, 1995, Orlando, Florida.

Lei, X., & Yang, D. (2024). Research on the impact of water-saving technologies on the agricultural production efficiency of high- quality farmers taking Jiangxi province and Guangdong province in China as examples. *Frontiers in Environmental Science, 12*, 1355579. https://doi.org/10.3389/fenvs.2024.1355579

Molden, D., Sakthivadivel, R., & Habib, Z. (2001). Basin-level use and Productivity of Water: Examples from South Asia. *IWMI Research Report* 49, Colombo: International Water Management Institute.

Namara, R., Nagar, R., & Upadhyay, B. (2007). Economics, adoption determinants, and impacts of micro-irrigation technologies: Empirical results from India. *Irrigation Science., 25*, 283–297.

Pandya, P., & Dwivedi, D.K. (2018). Constraints in adoption of drip irrigation. *Advances in Life Sciences. 5*.

Rogers, E. M. (1995). *The Diffusion of innovations* (4th ed.). New York.

UN. (2024). Water. Retrieved from: www.unwater.org (Access date: 02 July 2024).

Van Rijn, F., Bulte, E., & Adekunle, A. (2012). Social capital and agricultural innovation in Sub-Saharan Africa. *Agricultural Systems, 108*, 112–122.

Venot, J.P., Kuper, M., & Zwarteveen, M. (Eds.), (2017). Drip Irrigation for agriculture: Untold stories of efficiency, innovation and development (1st ed.). Routledge. https://doi.org/10.4324/9781315537146

Wang, Y., Wang, Z., Zhao, M., & Li, B. (2024). The influence of technology perceptions on farmers' water-saving irrigation technology adoption behavior in the North China Plain. Water Policy. 26, https://doi.org/10.2166/wp.2024.170

Woltering, L., Ibrahim, A., Pasternak, D., & Ndjeunga, J. (2011). The African Market Gar den: The development of a low-pressure drip irrigation system for smallholders in the Sudano Sahel. *Irrigation and Drainage, 60*, 613–621.

Zahoor, I., & Mushtaq, A. (2023). Water pollution from agricultural activities: A critical global review. *International Journal of Biological and Chemical Sciences, 23*(1), 164–176.

Chapter 3
Use of Technology in Agriculture: Some Examples in Turkiye and Other Countries

Orhan Kurt

Abstract Use of technology and/or new innovations in agricultural production has become important for both effective and efficient use of resources, saving time and labor. It plays, because, a critical role in increasing productivity, reducing environmental impacts, and achieving sustainability goals. In this chapter, some examples of the use of technology in agriculture are given, and also, some technology applications in Turkiye and other countries are discussed. An example of the application of digitization, use of sensors, and other automation equipment in agriculture based on the whole farmers of a village under the name of "Smart Village Project" in Turkiye for both crop and animal production and its components are explained. Similar examples from some countries in the world are also given.

Keywords Agriculture · Digitization · Sensors · Technology · Smart agriculture

3.1 Introduction

Agriculture is a very important issue today, and the fact that providing healthy food is becoming increasingly difficult. The importance of technology-supported agriculture is, thus, even more important both to increase production and to provide healthy food. Accordingly, the pioneers of the agricultural sector need to incorporate technology processes into their systems and keep pace with innovations in an ever-evolving global society in order to increase sustainability, efficiency, and profitability in production.

It is estimated that the world population, which was approximately 7.5 billion in 2016, will increase to 8.5 billion by 2030 and to 9.7 billion in 2050. In order to meet the food needs of this population growth, agricultural production must, thus, increase by 70% by 2050 (FAO, 2009). Thus, the global demand for food is increasing year by year depending on the growing global population. This situation, on the one hand,

O. Kurt (✉)
Agricultural Technologies Company, Istanbul, Türkiye
e-mail: kurtorhan00@gmail.com

© The Author(s), under exclusive license to Springer Nature Switzerland AG 2024   49
Ö. Çetin (ed.), *Agriculture and Water Management Under Climate Change*,
SpringerBriefs in Earth System Sciences, https://doi.org/10.1007/978-3-031-74307-8_3

has brought the agricultural sector back to the agenda as a strategic sector. Thus, it necessary to review the current situation and development trends of the agricultural sector all over the world. On the other hand, important natural events such as climate change and drought are already factors that affect directly agriculture.

In the thousands of years that have passed since the first agricultural activities, the methods and tools used in agriculture have constantly evolved. This evolution has gained momentum with the rapid advancement of technology today. The use of technology in agriculture plays a critical role in increasing productivity, reducing environmental impacts, and achieving sustainability goals. In addition, it has become important to increase productivity in agriculture and to do so by utilizing developing technology and innovative approaches since the amount of cultivated agricultural land will not increase significantly. Use of technology and/or new innovations in agricultural production has, because, become important for both effective and efficient use of resources and saving time and labor.

It was announced for the first time in Germany that an industrial cycle called Industry 4.0 will be entered, where information technologies and industry will come together, production will be carried out with maximum efficiency with integrated computer systems and artificial intelligence will come to the fore in 2011 and later (Enholm et al., 2022). It is of great importance that the agricultural sector is affected in this process. This integration means that the entire agricultural sector is connected to each other in real time and in constant contact. Agriculture 4.0 is defined as increasing efficiency and effectiveness in agricultural production with the use of information communication technologies in the agricultural sector. First of all, it creates access to reliable and healthy food, accelerates information sharing and decision-making processes.

Agriculture 4.0 applications are carried out by equipping agricultural machines and areas with sensors and communicating with each other, and it is aimed to increase productivity and quality with the use of modern technologies. Basic components of agriculture such as climatic and soil conditions, plant nutrition, irrigation and harvest times, disease and pest control which are very important for agricultural production, are presented to the information of the producers quickly and simultaneously, and effective use of resources is ensured using smart systems. The production costs can be significantly reduced by means of those technology-based applications, and the international competitiveness of countries is increased by producing quality products with high nutritional value. In addition, Agriculture 4.0 also points to environmentally friendly and sustainable agricultural production.

Studies have shown that the use of Technology-supported Agriculture increases productivity and profitability in agriculture (Nowak, 2021). Today, the technologies such as mechanization, walking robots, autonomous systems, and drones are used in almost all sectors. These technologies are also used in agricultural production processes and thus have very important advantages such as effective use of time and high profitability in making people's lives easier and reducing the use of manpower. Considering the above explanations, use of technology in agriculture becomes, thus, a necessity and if this problem is not solved with the support of technology, there are serious concerns about agriculture while this issue is a convenience in other sectors and an element that increases production efficiency, The development of the process related to this event is expected to rise at an ever-increasing circulation. Reaching healthy food and minimizing people's unnecessary labor work should be left to technology.

According to the "Smart Agriculture Market Research" conducted by Huawei in 2017, the value of the world smart agriculture market, which was 13.7 billion USD in 2015, is expected to increase to 26.8 billion USD in 2020. This means that the market will be valued twice in five years: In the CEMA (European Agricultural Machinery Association) "Agriculture 4.0: Future of Agriculture" report, it is stated that there are 4500 manufacturers producing 450 different agricultural machines with an annual turnover of 26 billion Euros in Europe and 135,000 people are employed in this sector. According to the same report, between 70 and 80% of new agricultural equipment sold in Europe has a precision agriculture technology component. Another point emphasized in the report is that smart agricultural practices will be the factor that will affect the agricultural sector the most by 2030 and will play a driving role in ensuring the sustainability of EU agriculture (Dryancour, 2017) (Fig. 3.1).

In addition, countries that realized the advantages of Agriculture 4.0 in a timely manner prioritized Agriculture 4.0 practices in both their national and common policies and accelerated their supports, incentives, and R&D studies in this field.

This chapter aims to present the developments and practices in this field to the readers with some examples from Turkiye and other some countries while emphasizing the importance of technology in agriculture. In addition, discussions on the policies for the use of agricultural technologies will help us understand the future potential of these technologies. The role of technology for the sustainability and productivity of agriculture is the main theme of this chapter. Thus, some examples of the use of technology in agriculture are given, and also some technology applications in Turkiye and other countries are discussed.

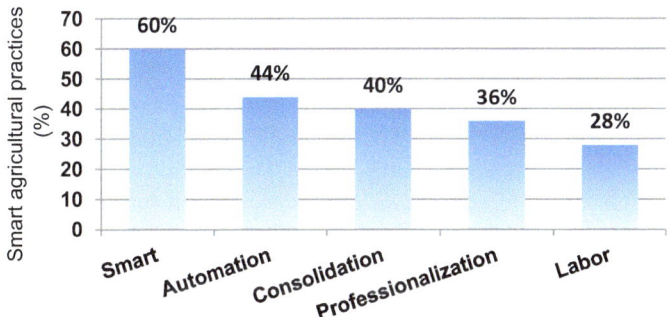

Fig. 3.1 The factors that will affect the agricultural sector by 2030

3.1.1   Use of Technology in Turkiye

Turkiye is an important agricultural country with its rich agricultural soils and climate diversity. In recent years, the importance of the use of technology in agriculture has gradually increased. In particular, innovative methods such as precision agriculture technologies, digital agricultural practices, and modern irrigation systems increase the productivity of Turkish farmers and ensure their competitiveness. For instance, precision agricultural technologies enable agricultural activities to be optimized by using tools such as soil and plant analysis, satellite imaging, drones, and global positioning system (GPS) technologies. For example, a study carried out in the Konya region of Turkiye showed that precision agriculture application significantly reduced water and fertilizer costs.

On the other hand, digital farming involves collecting and analyzing agricultural data through mobile applications and sensors. In Turkiye, digital platforms such as the Agricultural Information System (TARBIL), developed by the Ministry of Agriculture and Forestry, support decision-making processes by providing instant data and forecasting services to farmers.

Effective use of water resources is vital for the sustainability of agriculture in Turkiye. Modern irrigation techniques such as drip irrigation (surface drip and subsurface drip irrigation systems) and sprinkler systems, increase crop productivity by saving water (Çetin & Bilgel 2002; Çetin & Uzen 2016). The irrigation projects implemented in the Southeastern Anatolia Region of Turkiye are also among the successful examples in this field.

3.1.2   An Example Project for Use of Technology: Smart Village Project in Turkiye

Use of technology in agriculture will become widespread with the Smart Village Project which is the application area of Agriculture 4.0. The history of this project was started by Tulin Akın with the agricultural marketing e-commerce website (www.tarimsalpazarlama.com) in 2004 (Anonymous, 2024a). Tülin Akın has established a website and enabled farmers to access qualified information with this site. This information includes stock market prices, meteorological information, grants, supports, funds, diesel prices, etc. In addition to this information, the site was also used as a platform where a buyer could sell his products and meet with the seller. In 2004, due to the lack of widespread use of Internet and computers by farmers, the project evolved into the Vodafone Farmer Club project, a platform that allows this information to be sent to the farmer's pocket as an SMS. While the project was deemed worthy of many awards, it became a platform that served 1.5 million farmers, and farmers started to do technology-supported agriculture with the information they received from this service.

Considering the success of the project, they continued to explain the technology with a truck that would visit all the villages in Turkiye. They visited a total of 12,000 villages and tried to explain the technology to the farmers and spread the use of technology. However, digitalization and Agriculture 4.0 in agriculture have not gained a fast path and momentum as desired in Turkiye. Thereupon, Tülin Akın wanted to establish an exemplary village in Turkiye and started the planning phase.

The project planned should have been an exemplary study and should reflect all the villages. It should have been a center where all the villages could take an example and test the agricultural works in Turkiye. For this, they found the village of Kasaplar in the Koçarlı district of Aydın Province with a mathematical modeling. Kasaplar village is a village in Turkish standards and was a village with a population of 724, resistant to technology and a social structure representing the country.

Taking into account the characteristics of the land in this village, it was an area where 94% of agricultural products could be applied commercially in Turkiye. Many commercial crops are able to grow except tea, kiwi, hazelnuts, and bananas. At the same time, it is possible to grow three different crops at the same year under its climate and soil characteristics. This project started with the land purchased from the Municipality in the region, and it was based on the efficiency and profitability studies by applying the technologies on the agricultural lands.

This project components, technology applications, and devices within this Smart Village Project are briefly described as given below.

3.1.3 Traveling Hybrid Milking Unit

This project was the first in Turkiye and was designed because the goats and sheep in the pastures could not be milked. Due to the lack of electricity in the pastures and the lack of people to milk, goat and sheep milk is not obtained properly. The supply of this milk becomes difficult for people and the farmer could not benefit from the income generated from the milk production. Therefore, the device is designed with the support of General Directorate of Agricultural Researches and Policies (TAGEM).

The roaming hybrid milker can milk 24 sheep at the same time. It has panels that produce its own electricity, and thanks to these panels, milking is implemented without the need for electricity. Feeding units and milking are automatic, and when milking is over, the platform lifts and the animals get off the landing stops. The milked sheep milk is transmitted to the cooling tank and the system is designed to wash itself automatically at the end of milking.

3.1.4 Technology-Enabled Olive Production

In this system, data such as soil moisture, soil temperature, soil salinity, and leaf water content using sensors is obtained and all are transferred to the cloud data set. The water demand of the soil can be determined, the irrigation time can be planned, and supporting information can be produced for fertilizer studies by means of these data, Thus, the crops grow without stress, and the yield and profitability increase by means of an algorithm created,

3.1.5 Electronic Insect Trap

The farmers use pesticides for biological control and do not check whether there are insects and/or pests or not. It detects the moment the pest arrives and provides spraying if it is harmful using this system. The system works with a gel that emits an odor that attracts the insect in the box, and when the insect sticks, it takes a photo and transmits it to the cloud. The person who examines the data determines the insect population and establishes decision support mechanisms for spraying.

3.1.6 Early Warning System

The farmers can determine the irrigation time without going to the field and can operate the irrigation via mobile phone, tablet, and/or computer by means of this technology device.

3.1.7 Technology Supported Greenhouse

It is difficult to grow crops in non-technological greenhouses if some farmers want to use greenhouse. The important thing is to adjust the air circulation and temperature humidity control inside according to the crops. This is also to prevent the product from rotting. In fact, this is an important problem and that is why farmers avoid growing greenhouse crops. Soil moisture, soil temperature, air humidity, and air temperature in the greenhouse can be controlled and the corrected climate structure using those sensor,

3.1.8 Meteorological Station

The weather station is necessary to control the climatic conditions in agricultural lands and to carry out proper plant nutrition, irrigation, and fertilization and to follow diseases and pests, thus providing the opportunity to grow the appropriate crops. For instance, it can be precisely calculated the crop water consumption and/or requirement by analyzing the amount of precipitation.

3.1.9 Cattle Communal Milking Center

The most important problem of cattle farmers is the difficulty of milking. Due to the fact that milking is done every day in the morning and evening time, and therefore, the farmers must be at the farm all time, thus they give up keeping female animals. In this regard, a milking center was built in the center of the village, and the farmers were relieved of the difficulty of milking. Heat monitoring and disease follow-up for the animals are provided by means of the pedometers attached to their feet.

3.1.10 Tank-Well Automation

The farmers had to drill wells at long distances and have to wait long time in order to bring water to their fields. Then, when the existing tank overflows, they go and shut it down manually. When the float system reaches the lower point, the well trigger relay works and water begins to accumulate using this automation. When water reaches the upper buoy, the relay closes and thus the water accumulates automatically.

3.1.11 Beekeeping Automation

This system is designed to eliminate the disease and the risk of theft, which is a major problem of beekeepers. It can compare the in-hive performance with the weight of the hive through the sensors connected to the bottom of the hive, detect that there is a problem in the hive whose weight does not increase, and transmit a warning. When it lifts the trap of the hive or moves the hive, it sends a signal to the center and conveys that there is a risk of theft.

3.1.12   Smart Pasture System

The animals spend a long time of the day and need to graze in better conditions at the pasture for animal welfare. The welfare of the pasture is increased with scratching apparatus, cooling units, automation trough systems installed at the pasture. In addition, the smart door system closes the entrance of the animals to the pasture, where the disease is detected in the pedometers connected to the feet of the animals, and takes them to the quarantine zone. In this way, you can distinguish between healthy animals and diseased animals.

3.1.13   Technology Use in Different Countries in the World

Agricultural technologies are developing and spreading rapidly around the world. Indifferent countries, there are many innovative applications related to the use of agricultural technologies. Below are examples from some countries regarding the use of technology in agriculture.

3.1.14   United Kingdom (UK)

The UK is one of the countries that most successfully implements smart agricultural practices with the cooperation of universities, industry, and government sectors. It started the process by training young scientists and establishing research centers in this regard. The most important source of the UK's success in Agriculture 4.0 is shown as its support for agricultural research and training in this field. So much so that between 2011 and 2012 alone, the government spent 450 million Euros for research and development activities in the agriculture and food sectors. On the other hand, there are many institutions and organizations working in the field of agriculture in the UK. Among the main of these institutions is Ministry of Environment, Food and Rural Affairs (DEFRA), which works under the government. In 2016, DEFRA, Northern Ireland's Department for Economy, Environment and Rural Affairs, the Welsh Assembly Government, the Department of Rural Affairs, and the Scottish Government jointly published the "Agriculture in the UK" report, which describes agricultural data and the state of agriculture in the UK. There are quite striking data in this report. According to the report, they spent £250 million for agricultural technologies in 2015. In this way, wheat production increased from 7 tons per hectare to 8 tons. In 2015, income from the agricultural market amounted to £96 billion, equivalent to 0.7% of GDP. The people of 3.8 million are employed in the agriculture and food sectors. This accounts for 1.2% of the total workforce. One of the most important institutions working in the field of agriculture in England is the Rothamsted Institute. Celebrating its 175th anniversary in 2024, the main field of study of the institute is environmentally friendly agricultural technologies. The institute has carried out successful projects with a budget of 37 million pounds and 450 researchers. The cumulative contribution of their work to the UK economy is worth £3 billion annually. The researchers of the institution contribute to the dissemination of knowledge worldwide by publishing nearly 300 publications each year and offering 70% of the all free of charge. Currently, they are carrying out projects on food safety, genetic studies to increase productivity, agriculture, and food technologies of the future. In 2015, they established the world's first field crop analysis facility. The facility, which operates 24 h a day, has a scanner that can scan an area of 15×120 m with sensors and cameras on it. In this way, the development and health of the plants can be analyzed by scanning the field surface (Anonymous, 2024b).

In addition, one of the most important collaborations in the field of agriculture and food technologies in the UK is the N8 Agrifood platform, which is formed by eight of the country's leading universities. More than 450 researchers and more than 150 Ph.D. students work on this platform with a fund of 269 million Pounds. In addition, they also provide support to more than 40 businesses through their work. Their main areas of work are sustainable food production, safe food supply chain, and plant and food health. In this context, they focus on precision agriculture and Agriculture 4.0 technologies and gene studies.

As a result, the UK has become one of the leading and successful countries in the field of Agriculture 4.0 by emphasizing both national and international cooperation and attaching great importance to agricultural technology research and education.

3.1.15 Netherlands

Another example of a very successful country in the field of agricultural technologies is the Netherlands. Since half of the land is at least 1 m below sea level, nearly 60% of the country has been obtained by filling the lands under the sea. Despite this, the Netherlands ranks second in the world export of agricultural products. The Netherlands has 77% of the agricultural export volume of the European Union, 6% of the world fruit trade, and 16% of the world vegetable trade. It ranks 1st in the world with 8.1 billion Euros in tulip exports. Total agricultural exports reached a record 85 billion Euros in 2016. In addition, the Netherlands imports €4.6 billion worth of agricultural products from 107 countries and then packages and processes these products and exports them to more than 150 countries, earning €7.9 billion (Kumar & Fennema, 2017). The fact that three of the world's 25 largest food and beverage companies are located in the Netherlands and have a total of 4150 companies in the agri-food production sector is one of the important factors affecting this process.

The success of the Netherlands, which is one of the most advanced countries in the field of Agriculture 4.0 technologies, is due to its long-term and technology-based agricultural policies. The Dutch government has invested €1.4 million to purchase satellite data to improve the sustainability and productivity of agriculture, which provides farmers with information on soil, atmosphere and crop development online.

This collected data enables farmers to achieve greater efficiency and sustainability by closely monitoring the plants.

Behind the increase in productivity in the limited arable areas of the Netherlands, there is also the fact that it has been able to reflect its success in information technologies to agricultural technologies as well as agricultural policies. According to the data of the Netherlands Foreign Investment Agency (NFIA), 70% of the innovations in the Netherlands, which is the world's fourth largest exporter in the information services sector, are related to information technologies (NFIA, 2024). In this way, the Netherlands can both produce agricultural technologies and export these technologies. In 2015, the value of agricultural technology exports reached 9 billion Euros. Dutch producers; they have succeeded in increasing their production and efficiency with high-efficiency irrigation systems, advanced seed technologies, renewable energy systems, cobots and automation systems, big data analysis, and smart farm software.

Another advantage of the country is that in 2014, Wageningen University was ranked first in Europe and second in the world in the field of agriculture and forestry in the QS world university rankings. Wageningen University's economic research focuses on food safety, agriculture and food policies, monitoring of agriculture and food value chain with developed software, and production of agricultural technologies. Projects initiated and involved by the agricultural technologies working group include developing drones to automatically identify and map weeds, developing a gardening and harvesting robot, and producing durable and sensitive sensors.

It has an important share in these developments in the private sector in the Netherlands. With its eco-village project, Regen aims to establish villages in the Netherlands that produce their own food and energy. The first steps of the project, which aims to produce more organic foods by using technological infrastructure, to consume cleaner air and water, and to produce self-sufficient energy, were taken in the city of Amsterdam. Within this framework, an area with 25 houses is created 25 km from the city. The first houses are scheduled to be completed in the second quarter of 2019. The company has announced that if this project is successful, the project will be implemented in different European countries, especially Sweden, Denmark, and Norway.

The Dutch agricultural sector has a very strong international reputation, and the government supports this leading position by investing in innovation. In addition, universities, research institutions, agricultural food producing companies and technology producing companies make significant contributions to the agriculture and food sectors of the country in their studies in the field of agricultural technology.

3.1.16 United States of America (USA)

As the world's largest agricultural exporter, the secret to the success of the USA is its investments in both technology and teaching the use of technology. There are many institutes and sub-organizations affiliated with the US Federal Department of Agriculture (USDA). The aim of the National Institute of Agriculture and Food, which is one of them, is to increase productivity in agricultural production, to reduce food prices by reducing the use of water, fertilizers, and pesticides, to reduce the damage caused by agriculture to the environment and to ensure the production of safe foods. In this context, the institute supports research in physics, engineering, and computer science, agricultural tools, sensor and software production, and training for farmers on how to use technology.

In 2016, the "Select USA" summit organized by the US Department of Commerce emphasized the importance of agricultural production and agricultural technologies. Farmers, who are not satisfied with land reclamation and greenhouse cultivation for an increase in production, have started to use integrated systems that control temperature, humidity and pesticides online in their fields with the support of the state.

USA produces 80% of the world's almond production. However, almonds need a lot of water, so the production cost is high (Anonymous, 2016). In order to find a solution to this situation, moisture sensors were placed on almond trees and soil analyzes were made. The information collected in the cloud was transferred to the irrigation systems of the farms and irrigation was carried out appropriately. This technology has provided about 20% water savings in irrigation.

In addition, NASA sent an observation satellite into space to measure the amount of moisture in the soil. Every three days, the satellite transmits detailed information to scientists about droughts, floods and climate change.

In the progress of the USA in the field of agricultural technologies, private companies that produce both agricultural equipment and machinery and software within the framework of Agriculture 4.0 have a great role. In 2001, John Deere, the world's largest manufacturer of agricultural equipment, added GPS sensors to its tractors and other mobile machines, reducing the cost of fuel for fertilization and pesticides by nearly 40%. Many farmers have started using GPS for crop improvement and yield mapping. With the mobile application called Scoutpro, which was also developed in the USA, the producers can monitor their land live and can have detailed information of the conditions. In addition, Green Sense Ranch, located in Chicago, is the largest indoor vertical farm in the USA. High quality cops are produced by providing heat, humidity, sufficient amount of light and water with computer systems on the vertical shelves here. Since these crops are supported by creating a suitable environment with computer-controlled LEDs, crops can be taken twice a week.

The US Federal Department of Agriculture both provides incentives for integrated technologies for production and offers various support opportunities to farmers to use agricultural technology. With these incentives and supports, approximately, 300 billion dollars worth of agri-food products are produced annually in the USA.

3.1.17 Israel

Despite the fact that only 20% of Israel's land is arable due to its high salinity content, its natural water resources are below the United Nations' water poverty limit, and its agricultural labor force is quite low, it is now able to meet 95% of its food needs with its own production. In this context, Israel is perhaps the most striking example of successful countries. It has managed to turn all these disadvantages into advantages by means of the technologies it has developed and implemented. In addition to exporting vegetables and fruits worth approximately 2 billion USD every year, it also exports the fertilizers and agricultural technologies it produces to many countries. In order to reach this stage, first of all, the lands suitable for agriculture were rehabilitated. Vegetables and fruits produced from seven farms where has been established in the desert 150 m below sea level are exported.

Due to the fact that its land is desert, it exports 90% of the products it produces, although it cultivates only in the upper 30 cm part of the soil. Then, in order to find a solution to the irrigation problem, it started to treat salt water and wastewater in the industry and reuse it. In addition, 86% of the water used in irrigation in the country is provided by reclaimed water. Each solar panel placed can purify 3 thousand liters of salt water with the electrical energy produced daily. The temperature can be kept under control for 12 months with pipes laid under the ground. Many companies have been established that produce fertilizers using various technologies, especially the Israel Chemical Company, one of the largest fertilizer companies in the world.

Evogene, one of Israel's leading seed companies, strives to increase crop productivity through research and development in plant genetics and biotechnology. It has developed with the technology Afimilk provides producers with information about both the health status of the animals and the quality of the milk in real time. Eshet Eilon, on the other hand, uses X-rays with its spectral imaging machine to provide nutritional value, ripeness, quality information, and even when it will ripen. The company determines in advance a kind of mold fungus found in dates, which is the biggest obstacle to the date trade it has made to Arab countries, and helps to prevent the prevention of exports by intervening.

On the other hand, the Israeli government supports agricultural technologies, especially those for irrigation systems, biotechnology, and wastewater reuse. So much so that research and development expenditures in the field of agricultural technologies constitute 17% of Israel's budget. New technological start-up companies in the field of agricultural technology have a great impact on the transformation of the harsh conditions of Israel's agricultural sector.

3.1.18   Japan

The agricultural sector in Japan accounts for 1.5% of GDP. Japan, whose arable lands cover only 11% of its land, whose population working in agriculture is gradually decreasing. The average age of those currently working in the agricultural sector is quite high, and the products produced in rural areas are commercialized only in big cities and sometimes even in international markets. Japan has started to revitalize the agricultural sector by means of its investments in agricultural technologies. Considering agricultural technologies, on the one hand, increasing production, productivity, and quality, they have started to direct the attention of citizens to agriculture again by making agriculture attractive.

The Ministry of Agriculture, Forestry, and Fisheries is responsible for decisions related to agriculture in the country. According to the Annual Report on Food, Agriculture, and Rural Areas (2016) published by the Ministry, reducing input costs, implementing structural reforms in the distribution and processing process, and establishing a strategic export system have an important place in agricultural policies. Agricultural technologies are seen as the most important factor in reducing input costs.

Universities, technology centers and the private sector come to the fore in agricultural technologies. An area of 2000 m^2 was established for vegetable factory at Osaka Prefecture University. In this factory, it provides twice faster product development and harvests can be harvested 20 times a year by using only artificial lights without using any sunlight. These vegetables, which are produced sterile, can be consumed without washing. Tokyo University of Agriculture and Technology mainly conducts studies in the field of robotics. The wearable mechanical skeleton designed by scientists at the university makes life easier for farmers during the harvest of hand-picked products and enables faster harvesting. In this way, leg fatigue and pain of farmers were reduced by 50%, arm and shoulder fatigue and pain by 85%. In addition to technology production, the Fukushima Agricultural Technology Center provides technical support to local farmers, conducts awareness studies on the importance of agriculture, and provides free laboratory facilities for farmers to produce and use technology. The Spread company produces 10 million pieces of lettuce per year with robots on a farm. In this process, 98% of the water used on the farm is sent for recycling. Temperature, humidity, light, and carbon dioxide levels are adjusted by computers. By means of the hydrogel film designed with a thickness of 0.06 mm, Mebiol provides high efficiency in plant production with a very small amount of water. Water, necessary vitamins, and minerals are given to the plant, and harmful substances are filtered by means of this film. Pattruss company has succeeded in extending the shelf life of the products with the pyramid-shaped plastic packages it produces.

On the other hand, Japan, which is very successful in soilless agriculture, has managed to collect fruit in 8 s by accelerating production and harvesting by means of the robots. Producers in the bio-farms created; it can produce by controlling temperature, humidity and light with computers. In addition, by monitoring the temperature, daylight duration, and water holding capacity of the soil with cameras and sensors, they can fight pests and diseases and determine the right harvest time. Japan can produce 10,000 tomatoes by filtering harmful sun rays from a tomato plant with a rotating lens system and giving only beneficial rays. At the Tsukuba Science Expo, 16,897 tomatoes were obtained from 1 tomato plant with the same method.

Japan's agricultural exports increased by 24% and an income of 35 billion dollars were achieved using all those technologies. The Japanese government continues its price support policy to support farmers and make agriculture attractive. It also works to facilitate the commercialization of rural production.

3.2 Conclusion

For the successful implementation of agricultural technologies, effective and inclusive use policies are needed. These policies are important to ensure that the technology is widespread and that farmers benefit from these innovations. Many governments offer a variety of support programs and incentives to encourage the adoption of agricultural technologies. Financial support such as grant programs, low-interest loans, and tax breaks facilitate farmers' technology investments. In Turkiye, institutions such as KOSGEB and the Ministry of Agriculture and Forestry provide important support in this field.

Farmers need to be made aware and educated for the effective use of technology. Farmers' adoption and adaptation to technology should be ensured through agricultural consultancy services, training seminars and extension activities. In Turkiye, Chambers of Agriculture and Provincial Directorates of Agriculture undertake important tasks in this field.

In order for agricultural technologies to develop continuously and innovations to emerge, investments should be made in research and development activities. New solutions should be produced in the field of agricultural technologies with the cooperation of universities, research institutes, and the private sector. In Turkiye, Scientific and Techonological Research Council of Turkiye (TÜBİTAK) and universities carry out important projects in this field.

As a result, use of technology in agricultural production has become important for more efficient use of resources and less inputs, increasing yield and quality, saving time and labor. However, each technology requires a new field of study; it brings complex applications as well as the need for more information. Therefore, training and consultancy services should be provided at the right time and by the right people, and institutions and governments should support this.

References

Anonymous. (2024a). Permaculture plants: Clay pot olla irrigation. Retrieved from: https://permac ultureplants.com/olla-irrigation (Access date: 08 June 2024).

Anonymous. (2016). California almond industry facts. Retrieved from https://www.almonds.com/ sites/default/files/2016_almond_industry_factsheet.pdf (Access date: 11 June 2024).

Anonymous. (2024b). Rothamsted research. Retrieved from: https://www.rothamsted.ac.uk/sites/ default/files/Download/Rothamsted_Harpenden_Walks_Science_Trail.pdf (Access date: 17 June 2024).

Çetin, Ö., & Bilgel, L. (2002). Effects of different ırrigation methods on shedding and yield of cotton. *Agricultural Water Management, 54*(1), 1–15.

Çetin, Ö., & Üzen, N. (2016). Raising water productivity levels and ensuring sustainability of ırrigation for high water using crops. *2nd World Irrigation Forum*, 6–8 November 2016, Chiang Mai, Thailand. W.3.1.01

Dryancour, G. (2017). Smart agriculture for all farms. European Agricultural Machinery Association. Retrieved from https://www.cema-agri.org/ (Access date: 11 June 2024).

Enholm, I. M., Papagiannidis, E., Mikalef, P., & Krogstie, J. (2022). Artificial intelligence and business value: A literature review. *Information Systems Frontiers, 24*, 1709–1734. https://doi.org/10.1007/s10796-021-10186-w

FAO. (2009). Global agriculture towards 2050. Retrieved from https://www.fao.org/fileadmin/templates/wsfs/docs/Issues_papers/HLEF2050_Global_Agriculture.pdf (Access date: 8 July 2024).

Kumar, P., & Fennema, F. (2017). From potato eaters to world leaders in agriculture. Rerieved from https://blogs.worldbank.org/en/endpovertyinsouthasia/potato-eaters-world-leaders-agriculture (Access date: 14 July 2024).

NFIA. (2024). The Netherlands named one of the most innovative countries of 2023. Retrieved from: https://investinholland.com/news/the-netherlands-named-one-of-the-most-innovative-countries-of-2023/ (Access date: 9 June 2024).

Nowak, B. (2021). Precision agriculture: Where do we stand? A review of the adoption of precision agriculture technologies on field crops farms in developed countries. *Agricultural Research, 10*(4), 515–522. https://doi.org/10.1007/s40003-021-00539-x

Chapter 4
Irrigation Management and Innovative Approaches in Cotton Under Climate Change

Öner Çetin⊙

Abstract Cotton is primarily used as a raw material for textiles and garments but is also used for feed and energy. Climate change and water scarcity pose new challenges for cotton irrigation management. Since cotton is one of the most water-consuming crops, it can be irrigated with all irrigation methods if the necessary technical conditions are met. However, surface drip irrigation (SDI) and subsurface drip irrigation (SSDI) saved about 37–42% of water, respectively, compared to furrow irrigation. One of the ways to reduce water use and increase water efficiency in cotton is to use new technologies with pressurized irrigation systems and automation. This is based on some evaluation criteria such as irrigation system efficiency (%), distribution uniformity DU (%), total irrigation water used (m^3 ha^{-1}), water productivity (kg m^{-3}) and water economy, irrigation water economical productivity ($\$$ m^{-3}), and farmer's net income ($\$$ ha^{-1}). For this, besides use of the drip irrigation, use of some soil and plant-based innovative and technological methods and devices has become important in order to save water and increase water efficiency and productivity.

Keywords Climate · Cotton · Irrigation methods · Irrigation productivity · Water saving

4.1 Introduction

The world population is estimated to exceed 9 billion by 2050 (Kumar et al., 2024). The world's biggest challenges today are the demand and/or shortage of food for a growing population, adaptation to climate change, and environmental issues. The majority of freshwater on Earth is used for agriculture, daily life needs, and industry. Agriculture is a crucial driving force in both the food supply for growing populations and the economic growth of most countries. For this reason, sustainable agricultural practices have become increasingly important.

Ö. Çetin (✉)
Faculty of Agriculture, Dicle University, Diyarbakır, Türkiye
e-mail: onercetin@dicle.edu.tr

The basic components of agriculture are seed, soil, and climate. Human beings can be added to these as a factor that realizes and affects agriculture. Therefore, climate affects agricultural production almost directly. Soil and water resources, as natural resources used in agriculture, must be used efficiently for our future. In other words, it has become imperative to use the available natural resources for agricultural production in a sustainable manner without polluting and overconsuming all.

On the other hand, one of the most important environmental issues of recent years is global warming and climate change. Agriculture is the most affected and vulnerable sector, and its adaptation to climate change is correspondingly delayed and difficult. The fact that agriculture is extremely vulnerable to climate change means that increasing temperatures will reduce yields and increase the risk of disease and pest proliferation. Reductions or variability in precipitation could lead to reduced crop production, negatively affecting agriculture as a whole and increasing the risk to global food security. Regional cotton production will also be negatively affected.

Cotton is primarily the raw material of the textile and garment industry with its fiber; it is also an important industrial crop used for oil, feed, and energy. For this reason, it is a crop that makes a significant contribution to the national economy of most countries and plays an important role in agricultural employment and farmers' income. Approximately, 27 million tons of cotton fiber is produced annually in the world, and it is expected to reach 29 million tons by 2028, with an annual increase of 1.6% (Anonymous, 2024a; Hudson & Liu, 2024). More than 80% of the cotton produced in nearly 80 countries globally is produced in ten countries, including Turkiye. The top ten cotton countries in the world are China, India, the USA, Brazil, Australia, Turkiye, Pakistan, Uzbekistan, Argentina, and Mali (Fig. 4.1). China is the world's largest producer of cotton with 6.68 million tons of cotton production in the year of 2022–23 (Shahbandeh, 2023). As in some other countries, the need for cotton is increasing in parallel with the developments in the textile and garment sector, which occupies a very important place in production, employment, and exports and is an important part of industrialization and presence in global markets. Considering the world average, 1 kg of seed-cotton cost an average of $0.44. The countries producing 1 kg of seed-cotton at the highest cost are China ($0.75), Turkiye ($0.59), Bangladesh ($0.58), USA, and Greece ($0.56). Increasing cost and low sale prices observed in cotton agriculture in recent years have led countries to take measures to increase yield and reduce production costs. The countries that support cotton the most are China, the USA, India, and Turkiye (Özüdoğru, 2021).

Although rising costs and competition for resources (especially water) are important constraints, increased productivity due to technological advances will lead to a potential increase in cotton production in the near future (OECD-FAO, 2017). Moreover, cotton accounts for about 2.5% of the world's cultivated land, while 16% of the world's insecticides and 7% of the world's pesticides are used in cotton agriculture. In addition, annual CO_2 emission by cotton cultivation is about 220 million tons annually (Anonymous, 2024a; Hudson & Liu, 2024).

Water is critical for agriculture, and it is very difficult to grow cotton profitably without irrigation in most cotton-growing regions. Thus, cotton is one of the most water-intensive crops depending on the region. On the other hand, available water

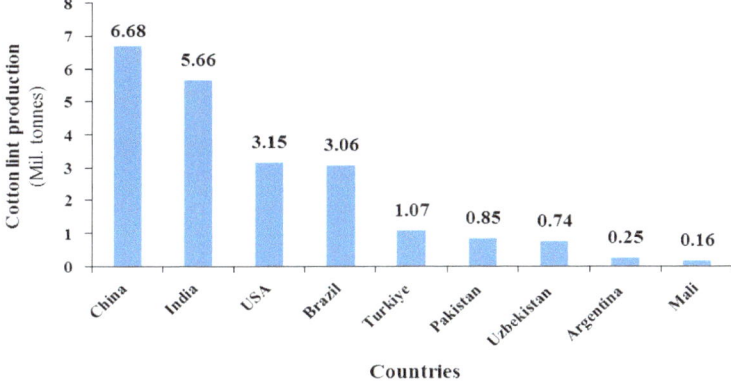

Fig. 4.1   Cotton lint production according to the countries (Shahbandeh, 2023)

resources will be adversely affected by global warming, and there is an urgent need to conserve water in agriculture. In addition, there is a need for more agricultural production for the growing population, and it should be taken into account sustainable practices by using water efficiently and economically.

Sustainable agricultural water management involves multifaceted and complex strategies such as developing a balanced management strategy according to the current water supply and demand, increasing irrigation efficiency and water productivity, wastewater use, and applying the necessary sanctions for differences in water use between basins. In addition, overuse of natural resources such as land and water should be avoided and utilized with protective measures. This can only be achieved through effective planning.

This chapter includes some main topics on the effects of climate change on cotton yield and production, irrigation relationship, irrigation scheduling that increases cotton yield and quality, the effect of different irrigation methods and/or systems on water use and water productivity, and the use of technology in cotton irrigation management.

4.2   Climate Change and Cotton Farming

Considering climate change, drought, and diminishing water resources, cotton production requires effective irrigation water management. Cotton (*Gossypium* sp.) is a crop that contributes significantly to the national economy in many countries and is used more than any other fiber in the world.

According to estimates, by 2080, climate change will cause a 20% increase in irrigation water demand and 10% decrease in agricultural production worldwide (Esteve et al., 2015; Fischer et al., 2007). Furthermore, according to the results of a quantitative climate change risk analysis, agricultural production losses due to climate change are expected to be 69%, 57%, and 45% at 90%, 50%, and 10% confidence levels, respectively (El-Nashar & Elyamany, 2022).

Regarding global warming, 14% of total greenhouse gas emissions come from agriculture, with methane (CH_4) contributing 52% and nitrous oxide (N_2O) 84% (Ton, 2011). Greenhouse gas emissions from agriculture are estimated to increase to 40% by 2030 (Smith et al., 2007).

Global climate change will definitely affect the cotton cultivation since cotton requires a certain level of moisture and temperature to produce sufficient yield and good-quality fiber. Increased CO_2 in the atmosphere increases photosynthesis up to a certain level, while extreme temperatures, on the contrary, have a decreasing effect on yield (Çetin & Başbağ 2010; Hughes, 2021). Some of these extreme weather conditions include extreme winds and temperatures, prolonged droughts, excessive rainfall, and floods, all of which can adversely affect cotton crop growth and productivity. According to a study, at least, one of the climatic risks will severely affect half of all cotton-growing areas by 2040 under a scenario of a global average temperature increase of 2.0 °C between 2045 and 2065, (Cunneen & Owain, 2021). However, it is argued that the impacts of climate changes on irrigated cotton are likely to be minimal, and rain-fed cotton will be more vulnerable (Jans et al., 2021). Therefore, it is inevitable for cotton producers to adapt to those extreme climatic events. Some measures that can be taken in this regard include soil and water conservation practices, use of newly developed varieties, reduction of synthetic agricultural inputs, and diversification of production (Farrel et al. 2023). Within the scope of climate change adaptation, changing the planting date of crops can also be considered as one of the options. According to Wu et al. (2023), the interaction of climate change and planting date had significant effects on cotton yield components such as temperature, light, and water use efficiency, but not on biomass. Therefore, it is clear that cotton growers need to take measures to minimize the impacts of global climate change, such as different cropping patterns and/or diversification of other practices (Cunneen & Owain, 2021).

It has been reported that the minimum temperature in the early period and the maximum temperature in the late period of cotton could cause significantly negative effect in cotton cultivation. The combination of temperature and rainfall or irrigation plays an important role in determining the length of cotton phenological periods. For a high level of statistical significance (>99%), it is mostly the combination of two variables, i.e., temperature (minimum, average, or maximum) and usually available water (rainfall plus irrigation) that determines the length of cotton phenological stages (Liakatas et al., 1998).

Climate change and water scarcity pose new challenges in cotton irrigation management. This is because climate change increases water demand and reduces crop yields. Therefore, it is important to identify and implement an economically optimal irrigation strategy. These are (i) irrigation strategies that ensure crop production at the lowest cost, (ii) irrigation strategies that enable more crop production per unit of water, and (iii) irrigation management strategies and value-added opinions and recommendations for decision makers and policy makers according to different crop species and varieties, different soil types and irrigation methods and/or systems (El-Nashar & Elyamany, 2022). Many cotton varieties have been developed to be tolerant to temperature, disease, and pests for local growing conditions. Therefore, the adoption of climate-resilient varieties offers an important opportunity for farmers to adapt to changing climatic conditions. For this, it is crucial that farmers in less developed countries have access to these varieties and that they should be tested according to their own conditions and receive support and training (Mandumbu et al., 2021).

On the other hand, although cotton agriculture is vulnerable to the impacts of climate change, it has the potential to mitigate climate change due to carbon retention in plant fibers and biomass. Forasmuch the cotton plant keeps 0.5 kg of CO_2 per kilogram of fiber produced (Hughes, 2021).

In addition, organic cotton cultivation has a lower carbon footprint, since synthetic fertilizers are not used and nitrogen oxides, a greenhouse gas, are not released. Cotton is more climate-friendly than most synthetic fibers used in the textile industry. It emits at least one-third less greenhouse gases per 1 kg of fiber produced. Growing cotton reduces greenhouse gas emissions toward the end user, from cotton production (5–10%), washing and drying (20–30%) and the highest (30–60%) during consumer use. Cotton is also readily biodegradable within 12 weeks, while synthetic fibers are not (Hughes, 2021). As a result, cotton is a fiber within the textile sector that clearly increases its potential to mitigate climate change and become more sustainable (Farrel et al. 2023).

Reducing agricultural water use while increasing productivity and total agricultural production requires the renewal and joint implementation of agronomic, engineering, and management systems as well as institutional alternatives. Only through these multistakeholder and multifaceted practices can this be achieved.

4.3 Irrigation Management in Cotton

4.3.1 Cotton Yield and Soil–Water Relationships

Water is an absolutely essential natural resource for all living things, agriculture, and the ecological environment. Managing water for all different users has become complex and challenging as climate change will also have significant impacts on water resources. Water has always been important for sustainability in cotton cultivation and industry. Higher cotton production per unit area can be achieved by improved irrigation systems and new agronomic practices without additional increase in water use or even with less water use compared to 20–30 years ago. Based on the irrigation water used volumetrically and both water productivity ($kg\ m^{-3}$) and economic water productivity ($\$\ m^{-3}$), all these contribute to a more careful and sustainable use of water resources.

Both excessive and insufficient soil moisture are not appropriate for growing cotton. This is because excess moisture causes more vegetative growth. During the vegetative growth period, excessive soil moisture also produces excessive canopy cover, which increases shading and restricts air movement. Thus, the opening of the bolls is delayed, and the time required for maturity increases. Adequate levels of available moisture in the soil prolong the flowering period, thus allowing more bolls to form as the total number of flowers produced will increase. During the flowering period, flower shedding increases significantly in plants exposed to moisture stress. Fiber yield is closely related to boll formation. Moisture stress at the stage of peak boll set significantly reduces yield. Under moisture stress at the boll development stage, the leaves can draw water from the bolls, and moisture stress lasting more than three days causes a significant yield reduction in cotton. Frequent irrigation delays boll formation and boll opening (Anonymous, 2024a).

Cotton is a taproot plant that grows rapidly reaching a depth of 20–25 cm in the initial period. Plant height is 20–25 cm and root depth can reach 90–150 cm. Most lateral roots are concentrated in the upper soil layer to a depth of 50 cm and can extend laterally up to 100 cm. Moisture stress restricts vegetative growth and promotes early flowering (Anonymous, 2024a).

A desirable fiber yield in cotton depends on maintaining an appropriate balance between vegetative growth (leaves) and boll (fruit) production. A planned moisture regime that does not adversely affect the yield too much and does not restrict vegetative growth is essential, especially during the first vegetative period of the plant. The cotton plant uses less water at the beginning of the growing period and more water is lost from the soil through evaporation than through transpiration. Water use in cotton increases in different regions from about 2–3 mm to a peak of 10–12 mm per day during the peak of flowering period (Fig. 4.2). It reaches its highest level during the boll formation period. If there is insufficient moisture in the soil, it should be irrigated before planting until the soil is at the appropriate moisture level. Adequate soil moisture is very important during flowering and boll formation. Moisture stress during these periods causes flower and boll drop, poor boll development, low ginning percentage, and ultimately low fiber yield. In addition, daytime sprinkler irrigation during extremely hot times also increases flower drop (shedding). The last irrigation should usually end when about 10% of the bolls have opened (Bilgel, 1995; Çetin et al., 2021; Anonymous, 2024a).

4.3.2   Critical Stages of Water Requirement and Response to Irrigation

A good irrigation management and agronomic practices as well as the use of an appropriate irrigation method/system significantly reduce the negative effects of drought in cotton. As with other crops, meteorological factors have significant effects on cotton growth and yield. During the cotton growth period, especially active radiation, effective temperature accumulation, and soil water content are very important. Nowadays, the application of irrigation water quantities targeting maximum yields

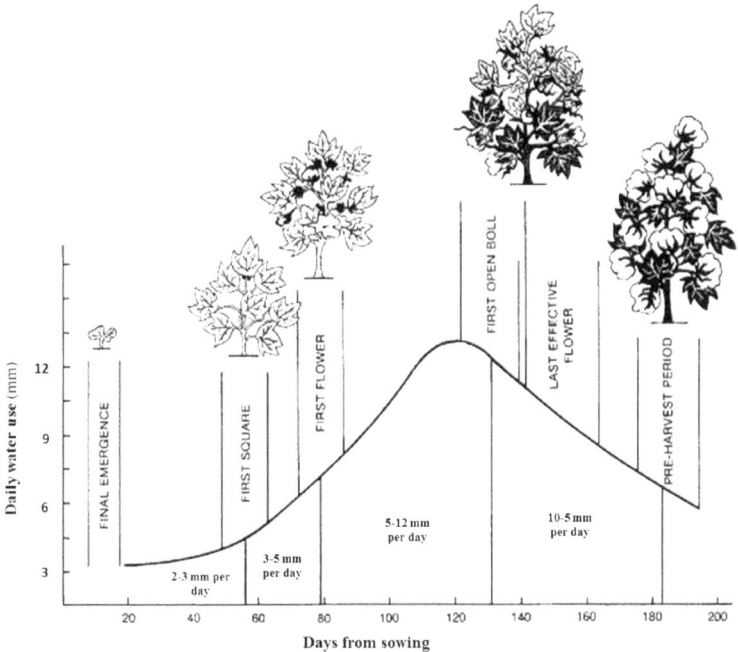

Fig. 4.2   Average daily water use between planting and harvesting of cotton cultivation

and covering the entire water consumption of the plant seems to be difficult due to diminishing and scarce water resources. Cotton is one of the most water-intensive crops in most regions. However, its water requirement is influenced by a wide range of factors, mainly climatic factors (temperature, precipitation (amount and distribution), relative humidity, wind speed, sunshine hours), crop variety, length of growing season, soil properties, and other agronomic practices.

Cotton accounts for about 20% of total seasonal water use up to the first flowering, 40% between the first flowering and last flowering, 30% between the last flowering and few bolls opening, and about 10% at maturity. Therefore, yield loss can be more than 50% if proper irrigation management described above is not practiced or if attention is not paid during water-sensitive periods (Datta et al., 2019). Therefore, higher cotton yields and water productivity should be targeted with the proper use of irrigation methods and/or systems that save irrigation water based on cotton development and water-sensitive periods.

Cotton is divided into three main periods in terms of water stress and/or water sensitivity. These are (i) vegetative development period, (ii) flowering period, and (iii) boll formation period. The most sensitive period to water stress is the flowering period. Accordingly, adequate soil moisture should be provided during the flowering period. During the flowering period, when soil moisture starts to decrease toward the permanent wilting point, boll retention rate decreases, and maturity may be delayed. Thus, significant reductions in yield occur. The second most sensitive period to water stress is the boll formation period. Therefore, keeping the soil profile at or near field capacity from early flowering to peak flowering will also support early maturity (Kara & Gündüz, 1998; Çetin & Bilgel 2002; Shekoofa & Raper, 2023). Considering these data, if the deficit irrigation water is to be applied in cotton, it should be done during the vegetative growth period which is less sensitive to water.

Apart from this, if there is a suitable and sufficient moisture in the soil for emergence during the cotton planting period, irrigation is not applied until about the first 30–40 days after the plant emergence, so that the plant develops a good root system and holds firmly to the soil. Excessive water application in cotton causes overdevelopment of the vegetative parts, and plant nutrients can be leached away by the effect of deep seepage. This can result in increasing fertilizer costs and causing groundwater pollution. Inadequate irrigation, on the other hand, causes moisture stress in plants leading to a decrease in the number of fruits (bolls) and poor boll development in cotton.

Knowledge of cotton water use, and plant water consumption (cop evapotranspiration, ETc) is very important for good irrigation planning and scheduling. Because plant water consumption (ETc) is an absolutely necessary data both in irrigation system planning and in making irrigation programs correctly. Plant water consumption is affected by climatic conditions (radiation, air temperature, humidity, wind speed), crop (crop type, variety, crop development stage), environmental and management conditions (soil salinity, inadequate nutrition, soil compaction, diseases and pests, cultivation and irrigation practices, windbreaks which reduce wind velocities across the adjacent field, irrigation systems that apply water directly to the root zone of crops (limiting evaporation losses as soil surface is dry), surface mulches which substantially reduce soil evaporation when crops are small). Numerous studies have been carried out on this topic (Ayars & Hutmacher, 1994; Farahani et al., 2008; Hunsaker, 1999; Kumar et al., 2015). The results showed that the water requirement (plant water consumption) varies between 700 and 1200 mm during the growing season, depending on season length, climate, crop variety, irrigation method, and production targets (Çetin & Bilgel 2002; Evett et al., 2012). For example, cotton irrigation water requirements for surface irrigation vary considerably even in different geographical regions of the same country (Turkiye) (Table 4.1).

Cotton water consumption and/or irrigation water requirement may vary considerably according to different climatic zones and different irrigation methods. For example, seasonal water consumption of cotton is around 730 mm in Australia, (Roth et al., 2012), whereas it can be much more than 1000 mm in the Southeastern Anatolia Region of Turkiye (Kara & Gündüz, 1998; Çetin & Bilgel, 2002). In recent years, significant increases in cotton yield and water productivity have been achieved. This can be attributed to advances in plant breeding, adoption of genetically modified varieties, and improved crop management. In addition, the use of irrigation scheduling tools and furrow irrigation system optimization and the use of modern irrigation methods/systems have also increased.

White (2007) determined that application of irrigation water at 79% of plant water consumption (ETc) resulted in a 31.5% increase in water productivity compared to 100% of ETc. Similarly, Yeates et al. (2007) reported that the highest yields using

Table 4.1 Amount of irrigation water and numbers for cotton irrigation under surface irrigation in different regions of Turkiye (Çetin, 2019)

Study area in Turkiye	Amount of irrigation water (mm)	Amount of irrigation water (m³ ha⁻¹)	Irrigation number	Sources
Iğdır	350	3500	3	(Istanbulluoğlu, 1995)
K.Maraş	650	6500	7	(Kanber et al., 1986)
Ege (Menemen)	450	4500	5–6	(Yalçuk & Özkara, 1984)
Ege (Nazilli)	350	3500	4	(Yalçuk & Özkara, 1984)
Çukurova	670	6700	5	(Kanber & Derviş, 1978)
Harran	900–1000	9000–10,000	8–10	(Çetin & Bilgel 2002)

drip irrigation were obtained using 75–83% of this, compared to 100% Class A Pan evaporation during flowering.

The studies carried out in different regions have shown that it is possible to reduce the amount of irrigation water used by cotton up to 50%, the cost of cotton processing and fertilization equipment. At the same time, it is also possible to increase cotton yield and water productivity and make more use of soil water through high-density planting (Çetin & Bilgel 2002; Orazgylyjov et al. 2023; Zuo et al., 2023).

For timing of cotton irrigation, some plant-based observations can be made before the appearance of a complete wilt due to lack of water. Plant leaves take on a slightly bluish, dull, and dull colored appearance. This can serve as a guide for irrigation. Irrigation should be done especially when wilting starts in the upper third of the plants. Irrigation in cotton can be based on climatic data as well as soil moisture monitoring. In this case, optimum irrigation can be performed when approximately 50% of the soil's available capacity, that is at the 50% of management allowed deficit (MAD), for surface irrigation. This rate may vary depending on the climate and the availability of water resources. For drip irrigation, irrigation should be carried out when 30–40% of MAD.

Changes in plant physiology, yield, and water productivity (WP) in response to water stress in cotton and the reasons for these changes are, in general, described by Gibb et al. (2012) as given in Table 4.2.

4.3.3 Irrigation Methods on Cotton

A good soil and management, accurate measurement of irrigation water applied to the land and improvement of metering devices, improvement of storage and water distribution systems, reduction of evaporation to drainage and evaporation losses, expansion of pressurized irrigation systems are important and necessary in order to

Table 4.2 Some responses of cotton to water stress (Gibb et al., 2012)

Degree of water stress	Possible causes	Physiological plant responses	Yield effects on maturity and WP
Minimal stress	Reduced irrigation deficit	Excessive vegetative growth	Reduced yield
	Excessive rainfall	Increase in leaf area	Reduced boll size
	Cloudy weather	Extended flowering cycle	Delayed maturity
	Excessive early insect damage	Reduced carbohydrate surplus for bolls	Normal fiber length but low micronaire
	High plant stands	Reduced root development	Poor WP
		High boll capacity but poor boll retention	
Mild stress	Optimum irrigation deficit	Optimum vegetative growth rate	Maximum yield
	Average temperatures (not excessively hot)	Leaf expansion restricted	High-quality cotton
		Photosynthesis remains unaffected	No delay in maturity
		Maximum carbohydrate surplus	Maximum WP
		Maximum boll development	
		Good fiber development	
Moderate stress	Increased irrigation deficit	Reduced vegetative growth and leaf expansion	Reduced yield
	Extremely hot temperatures with low humidity, windy conditions	Reduced photosynthesis	Early maturity
	Little cloud cover	Reduced surplus carbohydrates	Increased short fiber micronaire
		Reduced boll carrying capacity	Slight decrease in WP
		Increased fiber development	
Severe stress	Less than 3 irrigations	Vegetative growth greatly reduced–stops after flowering	Low yields
	Dryland crops	Greatly reduced carrying capacity	Short fiber

(continued)

Table 4.2 (continued)

Degree of water stress	Possible causes	Physiological plant responses	Yield effects on maturity and WP
		Little surplus carbohydrates	High or low micronaire depending on stress pattern
		Low boll retention	WP depends on rainfall

increase water productivity since excess water is used in cotton cultivation in most regions.

Cotton can be irrigated with all irrigation methods if the necessary technical conditions are met (Figs. 4.3 and 4.4). The important thing is to make good planning, projecting, and applications that take into account soil and crop properties by complying with the conditions required by each irrigation method. Each irrigation method can save irrigation water in itself. For example, it is possible to save 20–30% irrigation water in the most widely used furrow irrigation (Anonymous, 2024a). In other words, there is no significant decrease in yield by using an alternative furrow operating system. Furthermore, drip irrigation can save 30–50% irrigation water in cotton (Çetin & Bilgel 2002).

Fig. 4.3 Furrow (**a**) and sprinkler irrigation (**b**) in cotton cultivation

Fig. 4.4 Use of surface (**c**) and subsurface drip irrigation (**d**) in cotton cultivation

The use of pressurized irrigation systems (sprinkler, drip irrigation, low energy precision application (LEPA), moving drip, center pivot, or linear moving irrigation systems, etc.) increases year by year in cotton irrigation. This provides significant labor and water savings and is also suitable for automation. However, the use of these systems requires significant energy use, knowledge, and experience compared to surface irrigation.

On the other hand, although all irrigation methods can be used in cotton cultivation, each method and/or system has some advantages and/or disadvantages. For example, surface irrigation (furrow) uses more water than other irrigation methods because some surface runoff and/or deep seepage may occur. Sprinkler irrigation is also some advantageous than surface irrigation in terms of water application efficiency and water saving. However, it can save irrigation water in case it is used at the appropriate planning, system design, and irrigation scheduling. Otherwise, it may sometimes cause more irrigation water to be used than surface irrigation. Implementation of irrigation during daylight hours results in more water loss (10–15%) through evaporation, especially at high temperatures (>30 °C) during the irrigation season (Çetin & Başbağ 2010). In addition, water droplets falling on the plant leaves burn the leaves and/or flowers with a lens effect on the plant when irrigation is done at high temperatures during the daytime. It also significantly increases the rate of shedding compared to other surface and drip irrigation methods (Fig. 4.5). All of these can lead to a decrease in the number of bolls, and as a result, cotton yields may be lower than surface and/or drip irrigation methods. This explains why yields are lower with sprinkler irrigation. According to a study in three different irrigation methods, surface drip irrigation provided 21% higher seed-cotton yield than furrow irrigation and 30% higher than sprinkler irrigation (Çetin & Bilgel 2002). In addition, increasing water stress increased the shedding rate (Fig. 4.5) (Çetin & Bilgel 2002; Üzen et al., 2019).

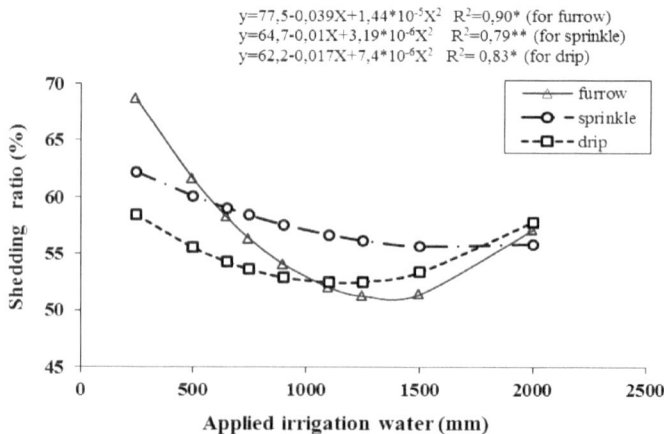

$y=77,5-0,039X+1,44*10^{-5}X^2$ $R^2=0,90*$ (for furrow)
$y=64,7-0,01X+3,19*10^{-6}X^2$ $R^2=0,79**$ (for sprinkle)
$y=62,2-0,017X+7,4*10^{-6}X^2$ $R^2=0,83*$ (for drip)

Fig. 4.5 Shedding ratio versus the applied irrigation water using different irrigation methods (Çetin and Bilgel 2002)

Use of modern irrigation methods such as surface drip (SDI) or subsurface drip irrigation (SSDI) is of more important in terms of water saving in regions with high temperature and low relative humidity, (Çetin & Kara 2019; Tarı et al., 2023). In a study, SDI and SSDI saved approximately 37% and 42% of water compared to furrow irrigation, respectively. In addition, net income per unit area was 20% and 69% higher under SDI and SSDI, than under furrow irrigation, respectively. Use of SSDI was more appropriate for irrigation scheduling based on the actual water consumption of the cotton crop ($1.0 \times ETc$) in saving water and maximizing irrigation water productivity ($0.80 \, kg \, m^{-3}$). Considering the possible water shortage, decreasing water resources, irrigation water savings, farmer conditions, regional and national incomes, the use of drip irrigation systems (especially SSDI) is extremely important (Cetin et al. 2021). It can be taken into account how water use, irrigation water productivity, and net income levels change if different irrigation methods/ systems are used for cotton-cultivated areas (Table 4.3). Of course, it is not expected that all areas will be irrigated with surface and/or subsurface drip irrigation systems. However, these data and scientific projections will be an important guide for farmers, irrigation authorities, and decision makers. It will even contribute to the governments' decision-making and development of policies to subsidize the farmers in case the farmers use drip irrigation systems. It should be noted that the seed-cotton yield, irrigation water productivity, water economic productivity, and total net income given here may vary from region to region and from farmer to farmer depending on many different factors mentioned in the previous sections.

In another study, it was determined that irrigation water (654 mm) at 95% of Class A Pan evaporation was appropriate for subsurface drip irrigation in a saline soil. Cotton yield decreased in irrigation water applications below this amount (DeTar, 2008).

Regarding some quality indicators of cotton, different amounts of irrigation water increased the fiber elongation and maturity index. Different drip irrigation systems have also affected fiber strength, and the maximum values were obtained from the subsurface drip irrigation (Üzen et al., 2019).

On the other hand, the percent wetted area and/or canopy cover is one of the most important criteria to be taken into account when applying irrigation water in drip irrigation systems. This is because the whole planted or cultivated area is not wetted or irrigated in drip irrigation systems. In arid and/or semi-arid regions with very hot temperatures (30–45 °C) and very low relative humidity (10–15%), it is recommended to be considered as 40% of the total water (i.e., 40% of total cultivated surface area) be applied or the area to be wetted at the beginning of the irrigation season for irrigation water calculation. Subsequently, the percentage of canopy cover used to calculate the amount of water applied should be based on this value (0.40) from the first irrigation until canopy cover exceeds 40%, and then total irrigation water should be calculated with the percentage of cover measured until the last irrigation. However, the canopy cover value used to calculate the amount of water applied has been suggested as an acceptable value of 80% (0.80) at the maturity stage, i.e., from the first boll opening to the last irrigation (about 2–3 weeks) (Cetin et al. 2021).

Table 4.3 Water use, irrigation water productivity, and net income in case of different irrigation methods in cotton-cultivated areas for Southeastern Anatolia Region of Turkiye¶ (Çetin, 2021)

Irrigation methods	Irri Water (m³ ha⁻¹)	Cotton cultivation area* (ha)	Seed-cotton yield (kg ha⁻¹)	Total production (Million Tones)	Total Irri. Water (Billion m³)	IWP** (kg m⁻³)	WEP ($ m⁻³)	Net income per area ($ ha⁻¹)	Total net income based on total area (Million $)	Total net income based on WEP (Million $)
	1	2	3	4 (2 × 3)	5 (1 × 2)	6 (3/1)	7 (8/1)	8	9 (2 × 8)	10 (5 × 7)
Furrow***	9500	293,000	3469	1.016	2.78	0.37	0.07	652	191	195
Surface drip	6000	293,000	4347	1.273	1.76	0.72	0.13	781	229	229
Sub-surface drip (40 cm)	5500	293,000	4383	1.284	1.61	0.80	0.20	1105	324	322

It is a projection study using scientific research results (TUBİTAK-1001, Project results registered 115,600; Çetin et al., 2021)

*Total cultivation cotton area in Southeastern Anatolia Region of Turkiye (TUIK, 2019)

IWP, irrigation water productivity; *: WEP, water economical productivity

****This data is taken from Çetin & Bilgel (2002)

4.4 Innovative Strategies in Cotton Irrigation Management

Agricultural irrigation significantly increases crop production. However, use of technology and know-how has an important issue for the sustainability of this increase and the efficient use of water and more water productivity (Fig. 4.6). Therefore, most developed countries have now started to use smart irrigation systems according to technological developments in order to use water effectively and efficiently in agriculture. The management of cotton and grain agriculture through a digital system of developing innovative technologies can lead to efficient use of resources and inputs, increase in yield and quality (Orazgylyjov et al. 2023).

Improvements on existing irrigation systems and the use of technology are crucial for increasing irrigation efficiencies and irrigation water productivity. For example, use of automatic irrigation systems plays an important role in saving and conserving water. This ensures that water as a natural resource is used more efficiently and that farmers understand this. At the same time, advanced systems allow for efficient water management as well as soil and weather monitoring. Therefore, today, the use of technological innovations has become indispensable for farms and agriculture. New technological approaches and devices can be used to achieve the required high-performance values that will ensure operational or farm efficiency in irrigation and its applications. Therefore, both the "Internet of Things" (IoT) and the use of sensor systems are essential for this. Furthermore, IoT provides monitoring and evaluation opportunities for irrigation management processes by reducing the total cost of technology use (Obaideen et al., 2022).

For example, IoT-based moisture and soil sensors are used to collect soil-related data. This data is stored in a centralized cloud. Features are selected by the "Completely Fair Scheduling" (CFS) algorithm. Clustering of the data is performed by the K-means algorithm. This will help to keep similar data together. Then the classification model is created using support vector machine (SVM), random forest, and Naïve Bayes algorithm. The model is trained, validated, and tested using the data obtained. Historical data on soil and moisture are also used in the training of the

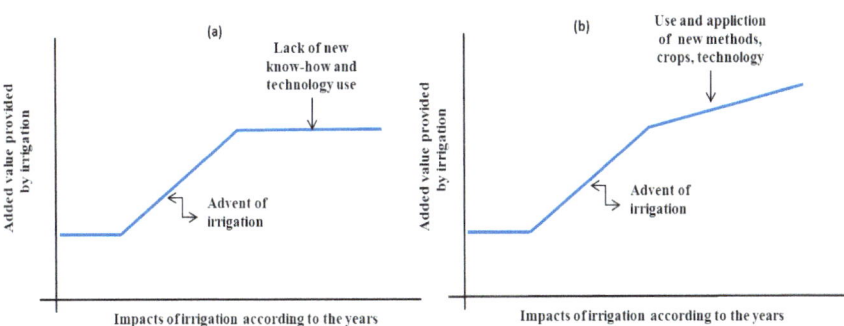

Fig. 4.6 Impact of irrigation on value added by years in case lack of new technologies and methods (**a**) and implementation of all (**b**)

model. K-means "support vector machine" (SVM) hybrid classifier achieves better results for classification, prediction of water demand, and freshwater conservation with smart irrigation (Kumar et al., 2024).

The right amount of water and nutrients can be applied automatically in real time by monitoring the distribution of moisture and nutrients in the soil using sensors. Thus, over- or under-application of water and fertilizer (fertigation) is avoided in the plant root zone. Accordingly, precision agriculture practices in cotton production have also emerged as an innovative approach as a key element for more sustainable cotton production. Precision agriculture is an agricultural management strategy based on observing, measuring, and acting on the agronomic and environmental situation, enabling farmers to be more precise in their use of resources according to prevailing soil and crop conditions. Precision agriculture involves modern technologies such as remote sensing, drones, and satellite imagery and enables farmers to monitor crop protection measures, soil moisture, and nutrient levels in the crops they grow with high accuracy. Thus, water losses, including runoff and evaporation, are significantly reduced. In addition, precision agriculture accurately predicts pest and disease outbreaks, enabling the adoption of integrated pest management (IPM) strategies (Anonymous, 2024b). As a result, precision agriculture helps to use less water, fertilizers, and pesticides.

On the other hand, remote sensing monitoring of crop production areas is becoming increasingly widespread all over the world. Crop coefficient (Kc) estimation based on spectral vegetation indices produced from multispectral images and determination of actual evapotranspiration (ETa) based on energy balance by combining multispectral, and thermal images with meteorological data make irrigation water management possible with remote sensing. ETa values determined in this way have great potential for both manual and automatic irrigation management. In particular, automatic management of fixed and mobile drip and sprinkler irrigation systems based on ETa maps offers a great opportunity to improve irrigation efficiency. Such irrigation management systems are also important for automatic variable rate irrigation (VRI).

The sophisticated systems are needed to help farmers increase production, reduce water wastage, and optimize product quality and management in the agro-industry during agricultural activities. Agriculture is highly dependent on climate and changing weather patterns which can lead to significant crop losses. One of the main problems is improper irrigation practices. Therefore, an innovative irrigation timing control system for smart agriculture could be useful for solving such problems or for sustainable resource utilization. Fuzzy logic can also be used to regulate irrigation scheduling in cotton crop fields, effectively preventing water wastage while ensuring that crops receive optimum water. For this, portable sensors measuring air humidity, temperature, and precipitation can be interfaced to collect real-time climate data. This climate data is sent to a fuzzy logic control system that dynamically adjusts irrigation scheduling and incorporates an algorithm that generates highly effective fuzzy logic rules and significantly increases irrigation efficiency by reducing total irrigation time. This automates the irrigation process and delivers the right amount of water precisely making irrigation management reliable without human intervention in the system (Bin et al., 2023).

In cotton irrigation management, whether modern irrigation technologies (drip, subsurface drip, smart irrigation, precision agriculture, etc.) or all other devices and new methods used or to be used, their advantages and effective use can be benefited if the necessary technical and application criteria are fulfilled.

4.4.1   Methods, Devices, and Sensors Used in Irrigation Management

4.4.1.1   Methods and Devices Used in Soil Moisture Measurement

Gravimetric method

(Weight Basis)

The basic measurement of soil water content is done by drying a weighted soil sample in an oven at 105 °C. Soil water content determined on the basis of weight alone does not make sense. Therefore, the wilting point and field capacity of soils should also be known. Directly measured soil water content is the basis of all methods and/or devices used to measure soil water content indirectly. In the fields where soil water is to be measured, for each method and device to be used, soil water is initially measured directly by gravimetric method and these devices are calibrated. Direct measurement of soil water content requires a large number of soil samples to be taken which is both time consuming and labor intensive (Çetin, 2003).

Measurement of Soil Water Pressure Potential

(Tensiometers)

Available soil water is retained by soil particles in the form of thin film strips. Under field capacity, the adhesion force between soil and water dominates and water cannot move independently. Under these conditions, water moves under a tension. Tension is inversely proportional to the amount of water in the soil and as the water in the soil increases, the movement of water becomes easier as the tension decreases. A tensiometer, a porous-ceramic piece of equipment, is used to detect or measure the retention energy (pressure) of soil water and the relationship between soil water content. Accordingly, in water-unsaturated conditions the soil water pressure energy is negative, often referred to as "tension." A tensiometer simply consists of a permeable (porous) ceramic tip in contact with the soil at the bottom, a tube filled with water, and a vacuum gauge at the top (Fig. 4.7). For irrigation programming, tensiometers can be used in two ways.

Fig. 4.7 A tensiometer and its parts

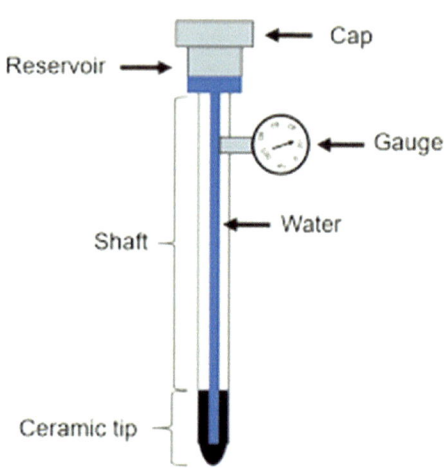

1. To apply a fixed amount of irrigation water whenever the tensiometer reading indicates stress conditions (threshold reading) (Alternating irrigation interval and fixed amount of irrigation water)
2. To irrigate at fixed intervals, applying varying amounts of irrigation water according to tensiometer readings (fixed irrigation interval and varying amount of irrigation water).

Tensiometers can be widely used in irrigation scheduling in field crops and orchards. Tensiometers are also used in the automation of irrigation. Accordingly, a capillary pressure threshold can be set, and the irrigation schedule can be started at this threshold. Thus, the irrigation system at a certain capacity can be automatically set the start and end times of the irrigation system by means of other devices that provide automation based on the amount of irrigation water or less amount of irrigation water that should be completed to the field capacity according to the predetermined threshold value (soil water pressure potential or soil water level read on the tensiometer) (Çetin, 2003).

In a study on cotton, soil water tension before irrigation was determined as 55 cb for surface drip irrigation if the tensiometer was placed at a depth of 15 cm and 47 cb if it was placed at a depth of 45 cm, respectively, in a very hot (>30 °C) and very low relative humidity (10–15%) region and soils which have more clayey texture. For subsurface drip irrigation, these values were determined as 52 cb if the tensiometer was placed at a depth of 15 cm and 45 cb if it was placed at a depth of 45 cm, respectively. It should be kept in mind that tensiometer threshold values may vary according to soil texture, irrigation method, whether or not to apply deficit irrigation, climatic conditions, and farmer decisions (Çetin & Üzen, 2018).

Neutron Scattering Method

(*Neutron Meter*)

Fig. 4.8 Schematic of a neutron probe SMC sensor deployed in the field (Kashyap & Kumar, 2021)

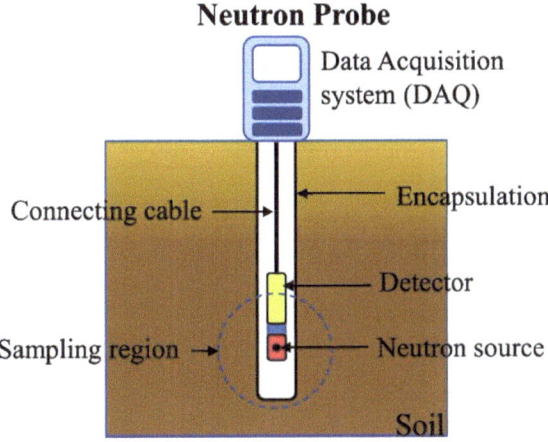

In the "neutron scattering method," a device called a neutron meter, consisting of a source that scatters fast neutrons and a detector that can count slowed neutrons, is used to measure soil water on a volumetric basis (Fig. 4.8). The most important advantage of a neutron meter is the instantaneous determination of soil water content. Neutron emission (scattering) occurs when an alpha particle (americium, plutonium, or radium) is mixed with tightly pressed Beryllium powder. In the soil, the atom that significantly slows down these neutrons, which are scattered at a very high speed, is hydrogen. In soil, the hydrogen form is normally found mostly in free water. When a fast neutron hits a hydrogen atom, the fast neutrons are slowed down and this thermal energy can be measured as slow neutrons. Thus, it is possible to measure the water present where it is desired to be measured (Çetin 2023).

Time Domain Reflectometry (TDR)

TDR is based on sending electromagnetic signals from a source through two or three probes placed into soil. These electromagnetic waves travel along the probes and return to the source as a reflection when the signals reach the soil through the probes and end up hitting it. The more water in the soil, the higher the dielectric power (slowing down the transmission speed more). It was first developed and used by Topp et al. (1980) that the propagation velocity of an electromagnetic wave can be measured using TDR and the motion (pulse) of the electromagnetic wave from the TDR device can be measured by a probe of L length inserted into the soil. Thereby, the soil water content can be determined based on volumetric unit. A TDR apparatus and probes are shown in Fig. 4.9A. The most important advantage of TDR is that soil volumetric water content can be measured directly. The disadvantages of TDR are that, besides being expensive, it does not give sufficiently accurate results and does not provide sufficient signal in soils which have high salty, organic matter, and high clay content.

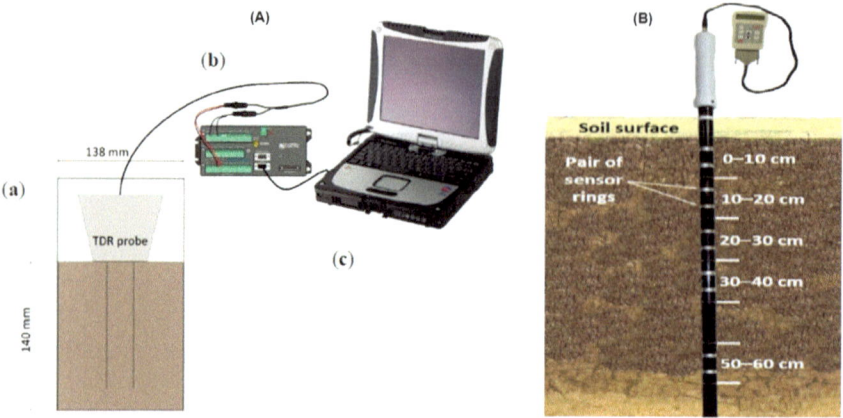

Fig. 4.9 Scheme of time-domain reflectometry (TDR) acquisition set-up; soil column with the TDR probe **a**; data-logger **b**, external PC for data processing **c** (A) (Vergnano et al., 2019) and profile probe (FDR) or capacitance sensors within the access tube (B) (Dhakal et al., 2019)

Frequency Domain Reflectometry (FDR)

(*Capacitance Sensors*)

Other sensors similar to TDR, that can measure soil volumetric water content are frequency domain reflectometry (FDR) devices based on the frequency of electromagnetic reflection. The working mechanism and physics are similar to TDR. Frequency domain sensors work in a similar way, but the frequency that occurs in a circular field surrounding an area or where the probe is located is used. The principles of FDR and TDR are the same, but the energy used is different. The approach in FDR is based on radio frequency waves to measure the capacity of the soil to hold electricity. Both TDR and FDR can be used in smart irrigation or irrigation automation as long as the conditions for their use are appropriate and the results are accurate (Fig. 4.9B). However, depending on the soil type, good calibration and accuracy must be ensured.

Remote Sensing (RS)

The basic principle of the methods used to estimate soil water by RS techniques is that water in the soil changes the surface temperature and affects the reflection changes in the soil. Changes in the intensity of this energy depend either on the dielectric strength of the surface or on the temperature of the surface, or a combination of the two. In remote sensing, satellite data, airplanes, and ground stations are also used, and only the soil water levels of the upper layer of the soil can be determined. Therefore, remote sensing data and processing results must be calibrated using ground real data (verification of actual soil water levels in the field with remote sensing data). Moreover, such methods are only practical for large areas and require expensive and sophisticated technology and specialized training and labor.

4.4.1.2 Plant-Based Measurements and Sensors

Some measurements in the plant, showing both the moisture level in the soil and the results of atmospheric effects, enable precise plant water status measurement for irrigation of crops. There are many different plant sensing tools for this. Satellites, aerial imaging systems, and hand-held instruments are often recommended and used to measure plant stress caused by water, nutrient deficiency and disease, and pests. In practice, however, there are a number of practical challenges in using plant-based sensing for irrigation scheduling. The development of inexpensive, wireless, and remote sensors has increased interest in the application of plant-based sensing techniques for irrigation scheduling.

Crop Canopy Temperatures

*(**Plant Spectral Sensors**)*

If soil water becomes limiting, the plant becomes stressed as water uptake is reduced and transpiration decreases, thus leaf temperature increases. When this is combined with the reduced amount of water in the soil, it causes the closed stomata that is the water conservation mechanism of plants. Therefore, leaves of well-watered plants are cooler, while leaves of plants under water stress are warmer on a hot and sunny day. Plant spectral sensors can be used to detect the properties (for example wavelength) of electromagnetic radiation (e.g., light) when it is reflected from a surface (Figs. 4.10a, b). This can be applied to detect water stress and nutritional status of the plant, i.e., the leaves, based on the reflection from the plant canopy. For example, plants under water stress show elevated canopy temperatures compared to fully irrigated plants. Under normal conditions, the plant opens its stomata, allowing water vapor to escape from the leaf due to atmospheric vapor pressure deficit and CO_2 to enter for photosynthesis. Water vapor escaping through the stomata cools the leaves (Roth et al., 2012).

Relative Water Content (RWC)

Leaf relative water content (RWC) is the ratio of the available water in sampled leaf tissues to the maximum water content it can hold at full turgidity. It is a measure of leaf water deficit. It is only used for research purposes under controlled conditions. Normal RWC values range from 98% in turgid and transpiring leaves to about 40% in severely dried and dying leaves. In most plants, leaf water content near wilting is about 60–70%.

Leaf Water Potential (LWP)

*(**Pressure Chamber**)*

A pressure chamber is used to measure leaf water pressure. For this purpose, the sampled leaves with the petiole are placed in the pressure chamber of the apparatus with the cut end of the petiole facing outwards, and pressure (nitrogen gas) is applied until water coming out of the cut petiole (Fig. 4.11a, b). The pressure value at the moment when water starts to come out from the petiole gives the leaf water potential value. This pressure is equivalent to the adsorption and capillarity forces of water and

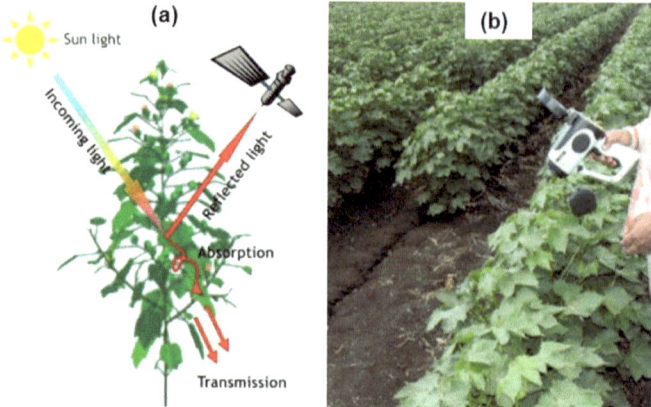

Fig. 4.10 Remote sensing consisting of object's illumination, light reflection and/or absorption, and light transmission through the atmosphere (**a**) and Infrared thermometer use on cotton irrigation management (**b**) (Ortiz et al., 2019)

the force of its adhesion to plant tissues. From a well-watered leaf with high water content, water outflow is easier (low pressure) while from a leaf under water stress, it is more difficult (high pressure), depending on the pressure applied. Leaf water potential values need to be correlated with soil water potential data (using neutron moisture meter and/or evapotranspiration) in order to use them as a tool for irrigation scheduling. It is possible to characterize irrigation scheduling for a given crop by means of this relationship.

According to a study carried out by Çetin (2020), the critical LWPs in vegetative period, flowering stage, and boll formation stage of cotton in surface drip irrigation for irrigation time were measured as −24, −23, and −24 bar, respectively. LWP is reference measuring of water status of cotton leaves and has enabled solid

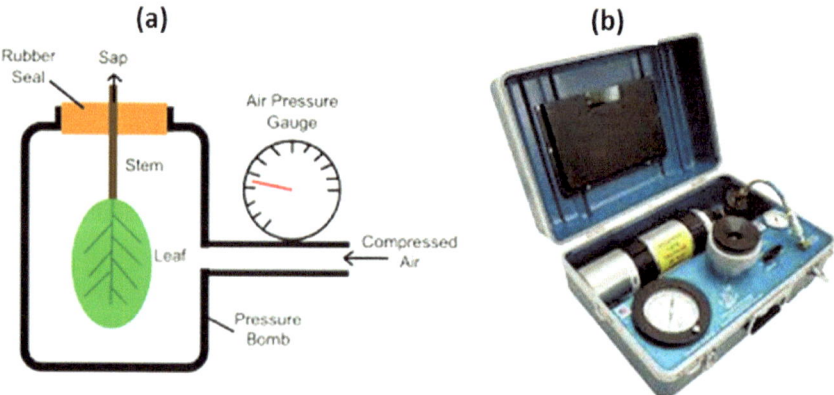

Fig. 4.11 Leaf water potential measurement mechanism (**a**) and pressure chamber device (**b**)

reference thresholds of cotton plant water status. However, LWP are highly affected by soil water status, light, air humidity, temperature, and calibration of the devices used. Therefore, these values as some threshold considerations may vary in different regions for the reasons mentioned.

Thermocouple Psychrometer

Measurement of leaf water potential by thermocapillary psychometry is based on measuring the temperature difference between the atmosphere and a freely evaporating moist surface of the leaves. The psychometer measures small differences in vapor pressure in the atmosphere (Fig. 4.12a). A thermocap is formed by joining two different metal wires (Anonymous, 2024c).

Stomatal Conductance (SC)

(*Porometer*)

Water loss in plants occurs in the leaves with the opening of stomata (pores), (i.e., transpiration) and CO_2 which is necessary for photosynthesis is taken into the plant tissue through the stomata. This is an important indicator for the physiological state of the plant. The opening of stomata can be interpreted as resistance to gas diffusion and can be measured using a porometer (Fig. 4.12b). Therefore, measurements of diffusion conductance in leaves provide important information on plant water status and also on plant growth and adaptation to environmental variables. Diffusion conductance in leaves decreases if the plant is under water stress and increases in well-watered plants. In a study, the critical SCs in vegetative period, flowering stage, and boll formation stage of cotton using surface drip irrigation were 312.8, 201.8, and 198.9 mmol m-^{2} s^{-1}, respectively. As in LWP, SC is also highly affected by temperature, soil water status, air humidity, light, and calibration of the devices used (Çetin, 2020).

Fig. 4.12 Use of a psychrometer Installed on small stem of a plant (**a**) the stomatal conductance porometer (**b**) (Anonymous, 2024c)

Fig. 4.13 Use of sap flow sensors in plants (Anonymous, 2023)

Sap Flow Sensors

Sap flow sensors are used to obtain information about the movement of sap in the plant. These can be categorized into three different groups. They are called heat balance, heat pulse, and thermal diffusion. The heat balance sensor is placed around the plant stem and the others require the placement of appropriate probes on the plant stems (Fig. 4.13). These sensors monitor changes in sap temperature when heat is applied to the stem and measure plant sap flow rate. Since transpiration in plants triggers sap flow within the plant, the measurements/data obtained can be related to the water status of the plant. However, these instruments are mostly used for scientific research.

4.5   Conclusion

Today, climate change is one of the biggest environmental problems. Due to this challenge and the increasing population, it is imperative to utilize the existing natural resources (soil and water) in the most effective and efficient ways. On the other hand, irrigation increases agricultural production significantly (1–5 times) depending on different regions and all other conditions (climate, crop type and variety, fertilization, soil properties, agronomic practices, etc.). Irrigation is the only way to achieve a significant increase in yield and production in general for cotton cultivation except for some individual regions.

On the other hand, further research and field experiments are needed to optimize the interactions of multiple factors (soil, water, agronomic practices, climate, greenhouse gas emissions, energy use, etc.) in cotton farming. Water productivity is highly variable among cotton farmers and farming areas and regions. Therefore, it is important to conduct site-specific measurements and research.

In cotton irrigation management, water consumption and the correct determination of irrigation water are important. In addition, the assessment of water productivity in cotton, the effect of different irrigation methods on cotton yield, deficit irrigation strategies, and the determination of the effects of plant density on yield and water use are among the important issues. Because one of the ways to reduce the cost of water is to increase water productivity in cotton cultivation. This will be possible through the use of pressurized irrigation systems that save irrigation water, the use of automation and new technologies in irrigation, and the breeding of cotton varieties resistant to drought or water deficit.

As a result, there are many different performance indicators that are taken into account as basis in irrigation management. These include irrigation system efficiency (%), distribution uniformity DU (%), total irrigation water used (m^3 ha^{-1}), water productivity (kg m^{-3}), water economic productivity ($\$$ m^{-3}) of irrigation water, and farmer net income ($\$$ ha^{-1}). All these indicators can be assessed on a farm, field and crop basis and are influenced by a wide range of factors. Therefore, all these indicators need to be identified to determine both sustainable and optimal irrigation practices and management at the field, farm, and irrigation scheme levels. In order to maximize and/or optimize these criteria, the use of drip irrigation, the use of some soil and plant-based innovative and technological methods and devices has become important in order to save water and increase water efficiency.

References

Anonymous (2023). Stomatal conductance. Environmental Biophysics, Measuring and Modeling the Environment Retrieved from https://environmentalbiophysics.org/tag/stomatal-conductance/ (Access date: 14 April 2023).

Anonymous (2024a). Cotton. Retrieved from https://www.ikisan.com/ap-cotton-water-management.htm (Access date: 2 February 2024).

Anonymous (2024b). Sustainability and precision agriculture in cotton production. Retrieved from https://trustuscotton.org/sustainability-and-precision-agriculture-in-cotton-production/ (Access date: 2 February 2024).

Anonymous (2024c). PSY1 Psychrometer (Stem & Leaf) for plant water potential. Retrieved from https://ictinternational.com/product/psy1-psychrometer-for-plant-water-potential/ (Access date: 18 April 2024).

Ayars, J. E., & Hutmacher, R. B. (1994). Crop coefficient for irrigating in the presence of groundwater. *Irrigation Science, 15*, 45–52

Bilgel, L. (1995). The first and last irrigation time for cotton in Harran Plain conditions. The Research Institute of Rural Affairs in Şanlıurfa of Turkiye. Publication No: 13/61, Şanlıurfa, Turkiye. (with English abstract in Turkish).

Bin, L., Shahzad, M., Khan, H., Bashir, M. M., Ullah, A., & Siddique, M. (2023). Sustainable smart agriculture farming for cotton crop: A fuzzy logic rule-based methodology. *Sustainability, 15*, 13874. https://doi.org/10.3390/su151813874

Çetin, Ö. (2003). Soil and water relationships and methods for measurement of soil moisture. The Ministry of Agriculture and Rural Affairs, General Directorate of Rural Affairs, *Research Institite Of Rural Affairs,* Publication No: 258/25, Eskişehir, Turkiye, p. 100. (in Turkish)

Çetin, Ö. (2019). Sustainable water saving and water productivity using different irrigation systems for cotton production. *3rd World Irrigation Forum,* 1–7 September 2019, Bali, Indonesia, ST-1-3, W.1.3.31

Çetin, Ö. (2020). Response of some physiological components of cotton to surface and subsurface drip irrigation using different irrigation water levels. *International Journal of Agriculture, Environment and Food Sciences, 4*(3), 244–254. https://doi.org/10.31015/jaefs.2020.3.2

Çetin, Ö. (2021). Agricultural irrigation and water saving in Southeastern Anatolia Region: Within the scope of Turkiye's strategy and action plan for combating agricultural drought. *Technical Report.* https://doi.org/10.13140/RG.2.2.19479.603251-6

Cetin, O., & Basbag, S. (2010). Effects of climatic factors on cotton production in semi-arid regions. *Research on Crops, 11*(3)

Cetin, O., & Bilgel, L. (2002). Effects of different irrigation methods on shedding and yield of cotton. *Agricultural Water Management, 54*(1), 1–15.

Çetin, Ö., & Üzen, N. (2018). Effects of surface and subsurface drip irrigation on soil water moisture variation and soil-water tension. *Journal of Harran Agricultural and Food Science (Harran Tarım Ve Gıda Bilimleri Dergisi)., 22*(4), 461–470. https://doi.org/10.29050/harranziraat.442314

Çetin, Ö., Üzen, N., Temiz, M. G., & Altunten, H. (2021). Improving cotton yield, water use and net income in different drip irrigation systems using real-time crop evapotranspiration. *Polish Journal of Environmental Studies., 30*(5), 4463–4474. https://doi.org/10.15244/pjoes/133238

Cunneen, H., & Owain, E. (2021). Physical climate risk for global cotton production: Global Analysis. *Forum For the Future, Cotton 2040.* http://www.acclimatise.uk.com/collaborations/cotton-2040/

Datta, A., Ullah, H., Ferdous, Z., Santiago-Arenas, R., & Attia, A. (2019). Water management in cotton. In: *Cotton Production, Chapter 3* (Eds. K. Jabran, B. S. Chauhan). https://doi.org/10.1002/9781119385523.ch3

DeTar, W. R. (2008). Yield and growth characteristics for cotton under various irrigation regimes on sandy soil. *Agricultural Water Management, 95*(1), 69–76. https://doi.org/10.1016/j.agwat.2007.08.009

Dhakal, M., West, C. P., Deb, S. K., Kharel, G., Glen, L., & Ritchie, G. L. (2019). Field calibration of PR2 capacitance probe in Pullman clay-loam soil of southern High Plains. *Agrosystems, Geosciences and Environment, 2,* 180043. https://doi.org/10.2134/age2018.10.0043

El-Nashar, W., & Elyamany, A. (2022). Managing risks of climate change on irrigation water in arid regions. *Water Resources Management.* https://doi.org/10.1007/s11269-022-03267-1

Esteve, P., Varela-Ortega, C., Blanco-Gutiérrez, I., & Downing, T. (2015). A hydro-economic model for the assessment of climate change impacts and adaptation in irrigated agriculture. *Ecological Economics, 120,* 49–58.

Evett, S. R., Baumhardt, R. L., Howell, T. A., Ibragimov, N. M., & Hunsaker, D. J. (2012). Cotton. Crop Yield Response to Water; *FAO Irrigation and Drainage Paper.* No. 66; FAO: Rome, Italy, pp. 152–161.

Farahani, H. J., Oweis, T. Y., & Izzi, G. (2008). Crop coefficient for drip-irrigated cotton in a Mediterranean environment. *Irrigation Science, 26,* 275–383.

Farrell, J., Larrea, C., & Luna, E. (2023). Global market report. Cotton prices and sustainability. *The International Institute for Sustainable Development,* 1–37. Retrieved from https://www.iisd.org/system/files/2023-01/2023-global-market-report-cotton.pdf. Access date February 9, February 2023

Fischer, G., Tubiello, F., Velthuizen, H., & Wiberg, D. (2007). Climate change impacts on irrigation water requirements: Effects of mitigation, 1990–2080. *Technology Forecasting and Social Change, 74*(7), 1083–1107.

Gibb, D., Constable, G., Neilsen, J. (2012). Cotton growth responses to water stress. In: WATERpak—a guide for irrigation management in cotton and grain farming systems (Ed. D.

Wigginton). *Cotton Research and Development Cooperation* (CRDC) 2012, Australia, ISBN 1 921025 16 6. pp. 239–247

Hudson, D., & Liu, B. (2024). Global cotton outlook, 2022/23–2031/32. International center for agricultural competitiveness department of agricultural and applied economics, Texas Tech University, Lubbock, TX 79409. Retrieved from https://www.depts.ttu.edu/aaec/icac/pubs/cotton/global_cotton_baselines/Final_Baseline_Mar2022.pdf (Access date: 7 February 2024).

Hughes, K. (2021). Cotton and climate change. Retrieved from www.icac.org. https://www.wto.org/english/tratop_e/agric_e/item_3_icac_climate_change__cotton_final.pdf (Access date: 16 April 2024).

Hunsaker, D. J. (1999). Basal crop coefficients and water for early maturity cotton. *Transactions of the ASAE 1999, 42,* 927–936

İstanbulluoğlu, A. (1995). Water consumptive use and last irrigation time of cotton in Iğdır Plain. *Research Institute of Rural Affairs in Erzurum. Publication No*: 49/45. (with English abstract in Turkish)

Jans, Y., Bloh, W., Schaphoff, S., & Müller, C. (2021). Global cotton production under climate change—Implications for yield and water consumption. *Hydrology Earth System Sciences, 25,* 2027–2044. https://doi.org/10.5194/hess-25-2027-2021

Kanber, R., & Derviş, Ö. (1978). Water consumptive use of cotton in Çukurova conditions. *Regional Research Institute of Soil-Water in Tarsus. Publication No*: 90/40. (with English abstract in Turkish)

Kanber, R., Eylen M., Yüksek, G., & Demiröz C. (1986). Water consumptive use of cotton under Kahramanmaraş conditions. *Research Institute of Rural Affairs in Tarsus. Publication No*: 113/63. (with English abstract in Turkish)

Kara, C., & Gündüz, M. (1998). Determination of the effect of deficit irrigation water on cotton yield in Harran Plain conditions of GAP Region. General Directorate of Rural Affairs, *Soil and Water Resources Research Branch, Publication No:* 106, Ankara, Turkiye. pp. 285–301

Kashyap, B., & Kumar, R. (2021). Sensing methodologies in agriculture for soil moisture and nutrient monitoring. *IEEE Access, 9,* 14095–14121. https://doi.org/10.1109/ACCESS.2021.3052478

Kumar, G. K., Bangare, M. L., Bangare, P. M., Kumar, C. R., Raj, R., Arias-Gonzáles, J. L., Omarov, B., & Mia, M. S. (2024). Internet of things sensors and support vector machine integrated intelligent irrigation system for agriculture industry. *Discover Sustainability, 5*(6)

Kumar, K., Udeigweb, T. K., Clawsonc, E. L., Rohlid, R. V., & Miller, D. K. (2015). Crop water use and stage-specific crop coefficients for irrigated cotton in the mid-south, United States. *Agricultural Water Management, 156,* 63–69

Liakatas, A., Roussopoulos, D., Angelakis, C., & Christodoulou, C. (1998). Effects of meteorological parameters and ırrigation on cotton phenology in Greece. *Proceedings of the World Cotton Research Conference-2.* Athens, Greece, September 6–12, 1998. pp. 469–475.

Mandumbu, R., Nyawenze, C., Rugare, J. T., Nyamadzawo, G., Parwada, C., & Tibugari, H. (2021). Tied ridges and better cotton breeds for climate change adaptation. In N. Oguge, D. Ayal, L. Adeleke, & I. da Silva. (Eds.), *African Handbook of Climate Change Adaptation.* Springer. https://doi.org/10.1007/978-3-030-45106-6_23

Obaideen, K., Yousef, B. A. A., AlMallahi, M. N., Yong Chai Tan, Y. C., Mahmoud, M., Jaber, H., & Ramadan, M. (2022). An overview of smart irrigation systems using IoT. *Energy Nexus, 7,* 100124

OECD-FAO (2017). Cotton in OECD-FAO. *Agricultural Outlook 2017–2026, OECD Publishing,* Paris. https://doi.org/10.1787/agr_outlook-2017-14-en

Orazgylyjov, B., Dowletov, H., Balyshev, T., Eziz Rejepov, E., & Hamydov, R. (2023). The use of smart systems in cotton irrigation as a way of water saving in the climatic conditions of the Dashoguz province. *BIO Web of Conferences*, 010, CIBTA-II-2023. https://doi.org/10.1051/bioconf/20237101091

Ortiz, B., Shaw, J., & Fulton, J. (2019). Basics of crop sensing. In Crop Production. *Alabama Cooperative Extension System*, ANR-1398

Özüdoğru, T. (2021). Cotton production economy in the World and Turkiye. UCTEA Chamber of Textile Engineers. *Journal of Textiles and Engineer*. 2021/2: 28 (122). (In Turkish).

Roth, G., Goyne, P., Brodrick, R., & Conaty, W. (2012). Plant water status measurement. In: WATERpak—a guide for irrigation management in cotton and grain farming systems (Ed. D. Wigginton). *Cotton Research and Development Corporation (CRDC)* 2012, Australia, ISBN 1 921025 16 6. pp. 180–187

Shahbandeh, M. (2023). Global cotton production 2022/2023, by country. Retrieved from https://www.statista.com/statistics/263055/cotton-production-worldwide-by-top-countries/ (Access date: 14 April 2024).

Shekoofa, A., & Raper, T. (2023). Cotton growth stages and water requirements. Retrieved from https://news.utcrops.com/2020/07/cotton-growth-stages-and-water-requirements/ (Access date: 15 March 2024).

Smith, P., Martino, D., Cai, Z., Gwary, D., Janzen, H., Kumar, P., McCarl, B., Ogle, S., O'Mara, F., Rice, C., Scholes, B., & Sirotenko, O. (2007). Agriculture. In: Climate Change 2007: Mitigation. Contribution of Working Group III to the Fourth Assessment Report of the Intergovernmental Panel on Climate Change [B. Metz, O. R. Davidson, P. R. Bosch, R. Dave, L. A. Meyer (eds.)], *Cambridge University Press, Cambridge*, United Kingdom and New York, NY, USA. pp. 498–540.

Tarı, A. F., Satış, F., & Akın, S. (2023). The effects of different irrigation levels and irrigation intervals on cotton cultivation: A study on yield, yield components, and fiber quality parameters. *Harran Tarım Ve Gıda Bilimleri Derg., 27*(3), 293–305. https://doi.org/10.29050/harranziraat.1323064

Ton, P. (2011). Cotton and climate change: Impacts and options to mitigate and adapt. *International Trade Centre*, Palais des Nations, 1211 Geneva 10, Switzerland.

Topp, G. C., Davis, J. L., & Annan, A. P. (1980). Electromagnetic determination of soil water content: Measurement in coaxial transmission lines. *Water Resources Research., 16*, 574–582.

TUIK (2019). Crop production statistics. Retrieved from https://data.tuik.gov.tr/Bulten/Index?p=Bitkisel-Uretim-Istatistikleri (Access date: 12 November 2019).

Üzen, N., Çetin Ö., Temiz, M.G., & Başbağ, S. (2019). Effects of different drip irrigation systems and irrigation management on cotton fiber yield, yield components and fiber quality. *Mediterranean Agricultural Sciences, 32*(3), 387–393. https://doi.org/10.29136/mediterranean.458025 (with English abstract in Turkish)

White, S. (2007) Partial rootzone drying and deficit irrigation in cotton for use under large mobile irrigation machines. *PhD Thesis*, University of Southern Queensland.

Wu, F., Guo, S., Huang, W., Han, Y., Wang, Z., Feng, L., Wang, G., Li, X., Lei, Y., Yang, B., Xiong, S., Zhi, X., Chen, J., Xin, M., Wang, Y., & Li, Y. (2023). Adaptation of cotton production to climate change by sowing date optimization and precision resource management. *Industrial Crops and Products, 203*(1), 117167, 1–13

Yalçuk, H., & Özkara, M. (1984). The effects of limited irrigation levels on cotton yield under the Aegean Region of Turkey. *Menemen Regional Research Institute of Soil and Water. Publication No:* 107/70. (with English abstract in Turkish).

Yeates, S., Strickland, G., Moulden, J., & Davies, A. (2007). NORpak-Ord River irrigation area. Cotton production and management guidelines for the Ord River Irrigation Area (ORIA) 2007. P38, *Cotton Catchment Communities Cooperative Research Centre.* http://web.cotton.crc.org.au/content/Industry/Publications/Northern_Production.aspx

Zuo, W., Wu, B., Wang, Y., Xu, S., Tian, J., Jiu, X., Dong, H., & Zhang, W. (2023). Optimal planting pattern of cotton is regulated by irrigation amount under mulch drip irrigation. *Frontiers in Plant Science, 14*, 1158329. https://doi.org/10.3389/fpls.2023.1158329

Chapter 5
Sustainable Use of Natural Resources, a New Requirement of the Route from Production to Consumption

Mioara Mihăilă◉ and Andy Felix Jităreanu

Abstract Natural resources constitute a topic of high interest in various fields of reference, including interdisciplinary ones. The consumption of natural resources and their value perception concerns both the economic and the political, social or administrative world. Agriculture, as the primary beneficiary and user of natural resources, is almost entirely dependent on the existence and quality of these resources. Since agriculture depends on satisfying the basic needs of individuals, it is extremely important to analyze resource consumption and production systems for which natural resources are the main inputs. Therefore, sustainable management of natural resources is among the most important issues. From the perspective of the production-consumption relationship and the sustainable development of consumption needs should be addressed simultaneously. In this sense, an eco-economic evaluation of a Natura 2000 site in Romania (Dobrina Forest) is presented, as an example and with the aim of emphasizing a sustainable way of assessing the consumption and exploitation of resources. The results indicate that ensuring sustainability on the production-consumption route significantly depends on the awareness of the vital role that natural resources have for the economic environment, as well as on changing the value perception of resource use.

Keywords Consumption · Natural resources · Needs · Production · Sustainability

5.1 Introduction

The use of natural resources is inevitable as long as life continues. The main reasons such as climate change, increasing population, industrialization and energy needs, and the increase in housing areas not only increase the use of natural resources

M. Mihăilă (✉) · A. F. Jităreanu
Faculty of Agriculture, Agroeconomics Department, Iaşi University of Life Science "Ion Ionescu de La Brad", Iaşi, Romania
e-mail: mioara.mihaila@iuls.ro

but also cause their rapid degradation. The importance of sustainable use of natural resources without causing further degradation has increased.

The requirements for sustainable development are increasing and that human needs are constantly dynamic and changing, both in terms of value and perception. But, a major alarm is being raised regarding the valuation and valorization of natural resources which, more and more, are seen only as inputs in production processes. The production-consumption path is a complex one, which is why it is necessary to reconsider the resources that are inputs for production and restore the value of these, under the conditions of sustainable development requirements. Building a framework for sustainable management of natural resources is a priority of today's world, in the conditions in which various crises—economic, social, ecological—are manifested at the global level. It is extremely important not to neglect or disregard the value of natural resources, without which no production system can run in optimal and sustainable conditions. To clarify the approach, it was used as an example of eco-economic calculation of the value of natural resources from a well-defined area, from Natura 2000 sites, precisely to highlight that it is possible to create various scenarios of approach, including for the sustainable use of natural resources.

The concepts that are the core of this material and for which a focused approach is made, on essential elements, theoretically and practically, are: the production-consumption relationship from the point of view of human needs, sustainability, adaptability, efficiency, and circular economy.

It is well known that economic and social systems cannot coexist and cannot progress without the consumption process, respectively the consumption process cannot be realized without the production process. Modern literature gives an extremely extensive space to the subject of "production" which wants to be efficient and abundant, as well as to the subject of "consumption" which ends up being a primordial element in market relations, in sales techniques and in marketing strategies. In fact, very few consumer entities, economic agents or individuals, pay more attention to the natural resources that are the basis of supporting the production-consumption circuit. The concept of sustainable development is increasingly invoked as a requirement to balance the link between production-consumption systems and the exploitation of natural resources.

The purpose of developing this material is to highlight the direct link of mutual influence between the issue of natural resources and that of sustainable development. The proposed objectives aim to show the production-consumption relationship from the perspective of reporting to the need to protect natural resources and demonstrate that without a primary consideration of the value of natural resources, both production and consumption, but especially the quality of life and general well-being can be altered and severely affected. This chapter addresses one of the most important problems of the global economy: the management of natural resources from the perspective of the production-consumption ratio.

5.2 Natural Resources and Consumption Patterns

The environment, with all the natural resources it includes, are the main components and source of the functioning of all the processes in the economic macro system. Some scientists and researchers are of the opinion that natural resources represent the "natural foundation of economic activities" that can favor or limit the development of human society (Herrmann-Pillath, 2013). Moreover, human being considered an integral part of nature, it is understood that the natural environment has a strong influential and determining role in the development of society.

Natural resources are viewed from several perspectives: a source of inputs for production processes, a natural "object" subject to appreciation and evaluation (including monetary), a decorative "object" that must be preserved. Concern for the state and quality of natural resources has acquired new nuances over the past 20 years or so, particularly in two different but related contexts: (i) international political and administrative requirements for sustainable development, (ii) the continuous and even alarming increase in the consumption of goods and products, an aspect that requires increasing inputs of resources. This is the general context in which the concept of sustainable development is inserted alongside the issue of natural resources. The 2030 Agenda of Sustainable Development Goals (SDG-17) represents a solution to concerns about the current state and future of natural resources (UN, 2023). The major current issues identified in relation to natural resources in sustainable development strategies, including the National Sustainable Development Strategy—Horizon 2030, are (UN, 2023): (i) improving the management of natural resources, (ii) avoiding their excessive exploitation, (iii) recognition of the value of services provided by ecosystems, and (iv) conservation and responsible management of natural resources. However, the natural resources are not only a topic of interest for policies, strategies or administrative concerns, but are also an object of study and objective analysis. As a simple definition, natural resources represent all the component elements of the environment, which can be used in human activity, regardless of its nature and purpose. According to Rajović & Bulatović (2017), from the point of view of sustainability or durability over time, natural resources are: (i) non-renewable: minerals and fossil fuels, (ii) renewable: water, air, soil, flora, wild fauna, and (iii) inexhaustible: solar, wind, geothermal, and wave energy.

Regardless of whether all are renewable or not, natural resources need attention, protection and proper management for at least the following reasons: maximum efficiency in use, productivity, solutions to their scarcity, quality assurance, monitoring, etc. Adequate management means balance and moderation, long-term thinking and decision-making, including consideration of future generations, overall vision.

Given that the topic of resources is extremely vast, complex and multi-targeted, in the present approach we will limit ourselves to only one category of resources, from the perspective of how they are treated as inputs for production processes, imminent predecessors of consumption ones. The resources we will refer to are the basic ones, which we also call strategic resources. The most relevant for the global economy, as well as the national one, are: lands, waters, forests, flora and fauna, biodiversity, and natural ecosystems. We will also refer specifically to the complexity of forestry resources in Romania, to Natura 2000 sites. The reasons for the customized approach are that these resources are extremely valuable on all three components of sustainable development (economic, social, and environmental) and that they are coveted by very heterogeneous entities not only at the national level, precisely because they are extremely appreciated and valuable.

The issue of natural resources is directly correlated with that of the environment, primarily through the physical belonging of the resources to the environment. In fact, all the resources that surround us and that humanity uses in the production and consumption processes, form the environment. Nowadays, the problems faced by the environment are no longer a concern only for specialists, but have become visible and alarming even for the general public. A prime example that we are witnessing, right now, are climate changes that reciprocally impact the management of natural resources. At least in Romania, these changes are clear evidence and directly affect the production sectors. Atypical phenomena, such as snow in April, storms, major and sudden changes in temperature, floods alternating with drought, it directly and negatively affects agricultural production processes, but also causes additional expenses for remediation of registered damages. Voices of public opinion claim that these are indirect consequences of hyper-consumption and aggressive exploitation of the environment, without considering strategies for its protection and conservation in a preventive manner.

The problems of production generating multiple losses and those of unsustainable consumption, analyzed along the production-consumption-production route, are current, they are real and, unfortunately, they are becoming more serious and more evident in almost all countries of the world. Confirmatory aspects are noticed worldwide, Europe is really worried, and in Romania the awareness and remedial efforts are increasing. The most common and urgent problems at the global level, from the perspective of an unsustainable course of resources on the production-consumption and return route, in the form of waste that constitutes a difficult problem from the point of view of their management, are the following (Sustainable Futures 2022):

1. *Production is inequitable from the perspective of resource management and reporting to the real needs of individuals:* Most of the time raw materials are exported from Romania (wood, cereals, salt, animal products, medicinal plants, etc.) and processed goods and products are imported, in the form of final consumption, which brings deficits in the trade balance, but also in the ecological balance of our country. Reports of a commercial nature indicate an ultra-abundance of consumer goods, which are in greater quantity than the needs that individuals have (examples: cars, clothing, shoes, some foods, etc.). Moreover, notes problems regarding production that increases in quantity but decreases in quality, which significantly affects the national budget.

2. *Urban expansion is detrimental to the protection and conservation of natural resources:* First of all, fertile agricultural lands are taken out of the production circuit, and the level of pollution and overcrowding is increasing. This situation is especially valid in developed or developing countries, the case of Romania

being a demonstrative one. Increasing areas of agricultural land have been transformed into peri-urban and metropolitan areas, support of urban and industrial infrastructure, creating agglomerations, and increasing pollution. Cities are the most affected by the dynamics of the current economy that operates according to a linear production-consumption-waste model. Although some of the towns-people migrate to rural areas for a better quality of life, this aspect makes the metropolitan areas expand more and more, without the natural environment being protected but, on the contrary, trampled (Felce & Perry, 1995). The solution that is increasingly discussed and already being implemented is the transition to a circular economy model.

3. *Industrial systems focused almost exclusively on profitability:* The tendency to accumulate capital and increase profit is constantly growing, and new modern technologies also have the role of developing and supporting industrialization, which is very necessary in a healthy economy. However, in order to relate to productive resources, not just to profit, it is necessary to rethink industrial systems, so as to align the requirements of sustainable development with those of healthy economic growth.

4. *Economic growth at an unbalanced rate between regions and economic sectors:* Sustained economic growth cannot be achieved without a high consumption of resources, a situation in which the resources of less developed areas, which do not have processing capacity and infrastructure, are used. This aspect creates and deepens social and economic inequalities between regions, in the sense that some regions develop at a very accelerated rate, while others lag far behind.

5. *Sustainable production becomes a competitive advantage, but the rate of transformation is slow:* Given the concentration of an increasing number of production units that put profit first, to the detriment of balanced growth, it has come to be that production units that assume sustainable production systems and that are gaining more credibility in the consumer market based on principles and values that consider not only profitability, but also the limitation of natural resources. That is why the transition from linear production systems, focused only on obtaining profit, to circular systems and responsible toward environmental factors, proves to be a guaranteed way of conquering the market or its stability, along with the contribution to reducing the effects adverse effects on the environment and available resources.

5.3 Degradation and Sustainable Management on Natural Resources

A major problem of current socio-economic systems, derived from production requirements and consumption needs, is the improvement of natural resource management systems. Interfering with this requirement for improvement are legal issues, stewardship and resource ownership, and issues related to the purpose for which environmental resources are used, sometimes brutally and aggressively. A relevant example is that of the exploitation of forests in Romania, shipped in massive quantities for export, which makes deforestation a high-level managerial-administrative concern and, most of the time, tense. Unfortunately, the benefits and ecosystem services of privately owned forests in Romania are disregarded.

In addition, the problem of reconsidering and activating a sustainable management of natural resources is raised both by the European Commission and by Romania, the priority being the problem of soils as support of agricultural production. 60% of these are assessed as unhealthy, the costs of the current degradation being estimated at 50 billion Euros annually. In this sense, the European Green Pact was developed, adopted on July 5, 2023, as a major form of international requirement that emphasizes the issue of sustainable use of natural resources (EC, 2019).

Accelerating the rate of economic growth by stimulating consumption, as an elementary principle of the functionality of economic systems, involved increasing the capacity and level of production in almost all sectors of economic activity. In simplistic logic, it is impossible to consume more if you do not produce more. An appropriate question in this context is: what is the purpose of the generalized orientation toward such accelerated economic growth? Gross Domestic Product (GDP) is appreciated as an indicator that reflects economic growth and most reports on GDP present the increase in its value as a successful result, without considering other indicators that could weigh the success of GDP, such as: quality of life, the state of health of the population, spending on health and environmental protection, the degree of pollution. The universally valid answer to the question "why permanent and accelerated economic growth is necessary" is related to the increase in human well-being, quality of life and standard of living. The economic use of resources, beyond the balanced utility analysis, turns into uselessness, imbalance, and involves much higher remedial costs. Examples are related to the invasion of real estate areas to the detriment of agricultural land, the clearing of forests in favor of the expansion of infrastructure, the allocation of resources for mainly commercial tourism purposes, etc.

5.4   Sustainable Development: Objectives and Requirements

Balanced production and sustainable consumption are not only ethical and moral aspirations, but have a solid economic and financial support, which involves improving the level of performance along the life cycle of a product, which in turn must be sustainable. A first step in rebuilding or adapting to a sustainable production system is always based on resources seen as inputs without which production cannot be achieved.

A new concept relevant to this context is that of smart consumption, which indicates the role of consumers in the protection and conservation of natural resources. Apart from individuals who, most of the time, allow themselves to be manipulated and influenced by various temptations to consume, whatever and however much, thus reaching hyper-consumption, an extremely important role is played by public institutions that can more proactively support sustainable consumption in primarily through honest and sustainable procurement (Angelova et al., 2021). "More is less and less is more" is a principle that indicates a balanced consumption of resources and satisfying only real consumption needs. In the modern, current economy, responsible consumption, and sustainable production have become fundamental and are the basis of decoupling economic growth from the problems faced by the environment and the pressure exerted on natural resources (EC, 2019). That is why a higher degree of awareness of the finite nature of resources and the application of measures that support objective 12 of sustainable development "responsible production and consumption" are requested. This objective requires economic agents, the political environment, researchers, but also consumers to adopt sustainable production and consumption practices, the circular economy being in the foreground as a concrete possibility to achieve the objective (de Jesus et al., 2021). The circular economy uses economic models that are based on the principles of the 10Rs: refuse, rethink, reduce, reuse, repair, restore, remanufacture, redirect, recycle, and recover.

Consequently, consumption behaviors must also become more conscious, only in this way is the rate of production corrected, which is important to become more efficient, not necessarily higher. At the international level, standards have also emerged that regulate these aspects. Thus, the ISO 20400:2017 standard (Sustainable procurement) provides guidelines for economic agents, regarding the integration of sustainability in the procurement process, a process that includes two components: the purchase of raw materials and production resources, respectively the purchase by the final consumer of finished products. Complementing this standard is the ISO 26000:2021 standard which regulates the requirements and implications of social responsibility at the organization level.

Another problem that arises in the analysis of progress in achieving the goals of sustainable development involving natural resources, production and consumption, is that of how to measure the stage of progress. In other words, how exactly is sustainable development measured and how do we know progress has been made or not? The answer is relatively simple and lies in indicator-based analysis. Since it was almost impossible to measure the progress of sustainable development in any other way, a system of indicators of sustainable development was established at the international level, one of the most representative office being Eurostat. The national systems have also adapted, and in Romania at the level of the National Institute of Statistics such a system was adopted and adapted. The only problem, for now, is that there is no harmonization between indicators at various hierarchical levels, which is why data collection is still difficult, but work is being done to remedy these aspects (UN, 2023). Then, it will be dealt with the presentation and analysis of a sub-system of indicators of sustainable development, in association with the main topic analyzed: the relationship between natural resources and the production-consumption chain, in terms of the requirements for achieving the objectives of sustainable development.

Responsible institutions in Romania, such as the National Institute of Statistics (NIS), relevant ministries, NGOs, associations, various authorities, have agreed, based on Eurostat data and other European recommendations, to use a complex and adequate system of indicators of sustainable development that includes the requirement to consider the complex value of natural resources. In Romania, NIS dealt with the development of a set of Sustainable Development Indicators, but is not entirely congruent with sustainable development indicators from Eurostat or other international authorities. According to NIS (2023), the main function of these indicators is to meet the requirements for monitoring the sustainability of the national economy, in accordance with the National Strategy for Sustainable Development—Horizon 2030, as an international political commitment (Anonymous, 2023). The indicators of sustainable development for Romania (NIS, 2018), numbering 103, are structured on three levels of interest and analysis, with time series available from the year 2000, as follows: (i) Level 1 indicators: they are the main or basic indicators (19 indicators); (ii) Level 2 indicators: are the complementary indicators, usable for monitoring and reviewing sustainable development programs (37 indicators); (iii) Level 3 indicators: are the progress indicators of the National Strategy for Sustainable Development of Romania and cover the package of European and international policies (47 indicators).

From the point of view of data content and significance, the national system of sustainable development indicators reflects the basic branches of sustainability: economy, society, and environment. In Table 5.1, the sustainable development indicators from the national statistical system are presented, those with which the state and progress of the natural resource heritage are monitored, respectively those that allow the analysis of the production-consumption relationship from the perspective of sustainability, are selected.

Productivity and efficiency in the use of production resources, which are extracted from the environment, are essential tools for a sustainable management of natural resources. Analytically, resource productivity refers to the resulting economic output per unit of material used. In Romania, according to NIS data, it decreased very little in the period 2008–2020 from 0.28 thousand USD/ton in 2008 to 0.27 thousand USD ton^{-1} in 2020 (i.e., 0.1%). During the analyzed period, there was a simultaneous increase in Domestic Consumption of Materials (DMC) and GDP at the national level, which indicates the tendency to increase the intensity of the use of materials and their productivity, as well as the decoupling of the economic system.

The material dependence of an economy is a modern indicator, highly significant for the progress analysis of sustainable development, and it measures the dependence of the economy on domestic natural resources. It also shows the extent to which an economy relies on the domestic extraction of resources to meet its material needs. In the period 2012–2020, in Romania there was a strong link of dependence of the economy on the domestic extraction of raw materials.

A synthesis of the dynamics in production and consumption, at the national level, based on the data recorded for the elaboration of the value of sustainable development indicators, indicates the following (Anonymous, 2023; NIS, 2023): (i) average total consumption expenses per household increased 2.2 times during the period 2008–2021, (ii) in 2020, the domestic consumption of biomass was 86.5 million tons, increasing by 64.9% compared to 2012, (iii) in 2021, the largest share in the total

Table 5.1 Sustainable development indicators representative of the sustainable management of natural resources, sustainable production and consumption (NIS, 2024)

Sustainable development objectives	Indicators (selection)	State of progress in Romania
O_2. Climate change and clean energy	Total greenhouse gas emissions (thousand tons)	From 117,875 in 2000, to 91,657 in 2018. Decrease by 22.25%
	Greenhouse gas emissions, by activity sector (thousand tons of CO_2 equivalent)	Energy: down 21.25%, from 97,787 in 2000 to 77,005 in 2018 Industrial processes: decrease by about 30%, from 19,187 in 2000 to 13,446 in 2018 Agriculture: 7.5% increase from 18,456 in 2000 to 19,854 in 2018
	Energy dependence (%)	Total: 32.6% in 2021 versus 22.7% in 2000 The biggest dependence is on crude oil: from 43% in 2000 to 67.9% in 2021
	Energy consumption per inhabitant (tep/inhabitant)	Increasing, in 2021 compared to 2000, from 1621 to 1783 (by about 10%)
	Final energy consumption by sector (thousand toe)	In 2021 compared to 2000 Industry and construction: down 24% from 9017 to 6849. Transportation: up 99% from 3508 in 2000 to 6976 in 2021. Population consumption: up 4% from 8433 in 2000 to 8765 in 2021. Agriculture and forestry: 43.5% increase, from 395 to 567. Other sectors: 172% increase, from 812 in 2000, to 2213 in 2021
O_4. Sustainable production and consumption	Productivity of resources (thousand lei in 2010 prices/ton)	Increase of about 47%, from 1.08 in 1995 to 1.59 in 2018
	Material consumption/person (tons/inhabitant)	50% increase from 13.50 in 1994 to 20.26 in 2018
	Municipal waste recycling rate (%)	Waste generated (tons): down 12.5%, from 6,040,230 in 2003 to 5,296,239 in 2018. Recycled waste (tons): up from 17,475 in 2003 to 586,406 in 2018 Waste recycling rate (%): increase from 0.29 in 2003 to 11.07 in 2018
	Share of organic crop areas in the utilized agricultural area (%)	Major growth to 3.8% in 2020 from 0.8% in 2006
	Electricity consumption in households (1000 toe)	Major increase of 86% from 658 in 2000 to 1226 in 2021

(continued)

Table 5.1 (continued)

Sustainable development objectives	Indicators (selection)	State of progress in Romania
	Enterprises with environmental management systems (No. of units)	Increase of about 20%, from 1181 in 2008 to 1414 in 2010
O$_5$. Conservation and management of natural resources	Area of sites of community importance (ha)	Increase by 41%, from 3,291,854.2 ha. in 2007, at 4,650,970 ha. in 2016
	Artificial space area as % of total area (%)	Increase by 12.5%, from 4.28 in 2000 to 4.82 in 2014
	Share of fresh water taken in total water resources (%)	Decrease of 30.5%, from 22.44% in 2000, to 15.6% in 2018
	Harvested woody mass 1000 m^3 (including bark)	37.5% increase, from 14,285 in 2000, to 19,652 in 2020

amount of imported waste was metal waste (45%), followed by paper waste (21%), mineral waste (18%), and plastic waste (14%), (iv) in 2021, the number of receiving operators who opted for reducing food waste increased 2.2 times, compared to 2019 (from 9 operators in 2019 to 20 in 2020), (v) the recycling rate of municipal waste in Romania was only 3.7% in 2020, (vi) the amount of waste generated per inhabitant decreased from 9.2 tons in 2008 to 7.3 tons in 2020 (-20%), (vii) the recycling rate of packaging waste increased from 33.5% in 2008 to 39.9% in 2020, and (viii) in 2020, the total amount of household waste collected was 4.8 million tons, increasing by 11.9% compared to 2010.

The indicators selected in Table 5.2. are relevant to the objectives of sustainable development, but there is a much better-defined set of indicators for Goal 12 of the global sustainable development strategy, "Responsible production and consumption". It is divided into two target objectives for the 2030 horizon, so it also has customized progress indicators (Table 5.2).

(1) O.12.1.: Phased transition to a new development model based on the rational and responsible use of resources and with the introduction of elements of the circular economy;
(2) O.12.2.: Halving per capita food waste at the retail and consumer levels and reduce food losses along production and supply chains, including post-harvest losses.

Table 5.2 Indicators for measuring the implementation of targets O.12.1 and O.12.2—Horizon 2030

Indicator	Progress	Value and argumentative dynamics
Main indicator		
Productivity of material resources	Red	Decrease from 1.28 thousand lei/ton in 2008 to 1.27 thousand lei/ton in 2020
Additional indicators		
Number of employees in the environmental goods and services sector	Red	Fluctuating values in the range 2014–2020, with a high of 168,000 people in 2014 and a low of 142,000 people in 2020 (decrease by 15.2%, while Bulgaria increased by 97.8%)
Share of GVA (Gross Value Added) from environmental technologies in GDP	Green	GVA in environmental technologies indicates their contribution to GDP; is the difference between the value of the production of the environmental sector and the intermediate consumption. In Romania, growth from 2.5% in 2008 to 7% in 2018, with an average annual growth rate of 11.44%
Share of renewable energy in gross final consumption of energy by sector	Yellow	In 2020 it was 24.50%, with 0.68 p.p. below the OECD target for 2030 (25.18%). In Romania, there are notable performances in the transport sector, in 2020, the share of renewable energy being 8.50%, with 7.2 p.p. more than in 2008
Material dependence	Red	2012–2020: very slow growth from 0.85 to 0.95%
Material intensity	Red	Increase from 0.52 tons/1000 lei in 2010 to 0.79 tons/1000 lei in 2020, which indicates the increase in the impact of the economy on the environment, from the perspective of the use of natural resources
The amount of waste generated	Yellow	2008–2020: oscillating evolution, with a maximum of 249 million tons in 2012 and a minimum of 141 million tons in 2020, that is, a decrease of 25.3% in 2020 compared to 2008 (from 189 million tons to 141 million tons)
Average total consumption expenditure per household	Yellow	As most of the expenses of Romanian households are allocated to consumption. In 2021, food products and non-alcoholic beverages accounted for 33.4% of consumption, down by 1.2 p.p. compared to 2020 and by 7.5 p.p. compared to 2008. The average annual growth rate during 2008–2021 was 4.54% Another component with an important share is the expenditure on housing, water, electricity, gas and other fuels (15.7%), increasing by 2.2 in 2021 compared to 2008 There are significant differences between the urban and rural environments, due to the consumption model, the level and structure of incomes

(continued)

Table 5.2 (continued)

Indicator	Progress	Value and argumentative dynamics
Internal consumption of materials, of which biomass	Green	Biomass consumption increased by 64.9% in Romania in 2020 compared to 2012, compared to 10.9% in Bulgaria, 36.7% in Hungary and 3.6% at EU level
Number of receiver operators who have opted for reducing food waste	Yellow	The reference period is short (3 years), because the implementation of the legislation in force only started in 2019. Increase from 9 operators in 2019 to 20 in 2021

Red = unfavorable situation, indicator in difficulty, Yellow = acceptable situation, but minor progress, Green = favorable situation, indicator in positive dynamics

5.5 Eco-Economic and Sustainable Utilization of Natural Resources

Another practical aspect of the analysis of the production-consumption relationship under sustainability conditions is that of the impact of resource consumption by impact categories, such as: climate change, resource consumption by category, and degree of pollution. Economic production units specifically use this type of analysis, including with the aim of obtaining the most advantageous products in various directions: the consumer market, competition, environmental protection, and product image. Other functionalities of the impact analysis are: waste management, evaluation of the benefits of avoiding waste production, evaluation of environmental management systems, sustainable economic evaluation of natural resources, value evaluation of the direct effects of production processes on the environment, benefits of waste recycling, etc. The transition from linear production-consumption systems to circular ones, involves numerous components that must function properly.

Ensuring the circularity of the production-consumption chain in the context of sustainable development is carried out according to rigorously determined norms, principles and standards. One of the most important is the analysis of the complex life cycle as an internationally standardized method through the ISO 14040 and 14,044 standards (ISO, 2015, 2024). This set of standards covers both the technical aspects of the production-consumption life cycle chain, as well as the managerial, organizational, administrative, and methodological ones.

Life cycle assessment quantifies all inputs of natural resources and energy in production processes and all outputs in the form of products intended for consumption and waste or emissions into air, water, and soil. All inputs and outputs are visualized and analyzed in a sustainability balance sheet, also called a dashboard of the production-consumption cycle. The values recorded in the scoreboard, as well as the sustainability balance sheet, are transposed into indicators precisely to obtain quantifiable values that can be analyzed. Thus, the pressure on the environment, the ecological footprint, the degree of depletion of resources, the alteration of flora and fauna, the impact of the economy and society on environmental factors, etc. are monitored.

Specifically, the notion of value of natural resources or nature in general can be interpreted in different ways. For an eco-economic analysis of the value, the notion is mainly used in relation to the usefulness of the analysis tools, to the promoted policies and strategies, to the national interests, etc. and much less in relation to human values, beliefs, principles, or standards. The use of the term "value" directly correlates with the intrinsic motivation of people to value and appreciate nature, to support efforts for its conservation and to consume in such a way that future production processes depend on them. The categories of values most used in the eco-economic analyzes of natural resources, environmental factors, or any natural area are presented in Table 5.3.

Another variant of the classification of the values that can be determined for the goods and services provided by the environment through the exploitation of natural resources, is proposed by Michieli (1993), who makes the classification according to the characteristics of the goods subject to the value analysis or the situation in which the evaluation of the respective asset is requested. Similar in content to the values presented above, the categories of values proposed by Michielli (1993) are: market value, production and/or reproduction value, cost value, transformation value, complementary value, replacement value, opportunity value, and social value. The value analysis of a good or service that belongs to the environment, also involves the stage of assigning a monetary value, i.e., a measurable amount of money to that good, service, or resource. Monetary valuation is, in essence, a procedure for estimating the value of the good or service in question, a value that is not constant over time, but varies under the influence or change of the conditions that determined it. The monetary value thus obtained is easily transposed into the systems of use and consumption in the economic sectors and in the social environment. Therefore, the problem of assigning value to natural resources is difficult in the context where the resources are used in very varied areas of interest, are rare or disputed as to ownership, etc.

A common denominator of resource evaluation remains valid: the efficiency of their use and the assignment of objective value. Given that the common interests of production systems with those of sustainable resource management are of the "win–win" type, the value analysis of the environmental factors used in the complex production-consumption process is limited to the following key points: increasing economic efficiency, improving responsibility of producers and consumers, reference to good sustainability practices. In this context, the solutions to create functional sustainable systems along the production-consumption route comply with specific steps, which bring benefits including in the short term (Table 5.4).

Efforts to implement sustainable development plans and strategies regarding natural resources valued as inputs for production processes have not remained without an echo. Both at the international level and in Romania, numerous projects have been

Table 5.3 Types of appreciable values in the eco-economic evaluation (ERDF, 2024; Mac Carthy & Morling, 2014; Strunz, 2014)

The value type	Meaning and specificity of assessment
Socio-cultural value	For most individuals, natural resources, biodiversity and environmental factors represent very important sources of well-being, primarily through the direct influence on mental health, through cultural values and traditions, but also other historical, national, ethical, religious, and spiritual values Socio-cultural values can be identified and quantified relatively simply, through direct methods and economic evaluation techniques, especially the transport cost method. Some resources are already protected precisely because they have been confirmed as essential for the identity and physical and valuable existence of individuals
Economic value	Most of the time it is the main value targeted by economic and financial-accounting systems regarding the material benefits and services provided by all exploitable resources in the environment (meadows, water gloss, wood, flora, fauna, landscapes) This type of value is valued strictly in monetary terms that indicate the effective contribution to human well-being and economic growth, being also found in the form of total economic value (TEV). It classifies goods according to how they are used in the production-consumption system
Use value	It is associated with the use of resources and goods exploited from the environment, by introduction into the production process and by direct consumption (example: consumption from crops, irrigation, hunting, etc.), direct non-consumptive use (such as the recreational use of resources natural) or indirect use (e.g., benefits associated with the flood risk reduction service generated by forest ecosystems)
Non-use value	It is associated with goods and resources that are not used, either directly or indirectly, in current production-consumption processes. Some goods can be appreciated and valued by their mere existence, even if there is no use in the productive system. A concrete example is that of the value that some people attribute to the existence and maintenance of a plant species in the field (especially if it is a rare plant, such as the cornflower in Romania). Non-use value can derive from the simple belief that the good or service in question must be available for use by other entities in current and future generations. This is where the altruistic or inherited value derives. Non-use values are almost always much more difficult to estimate than use values
Optional value	It refers to the value associated with delaying the use of a good or natural resource in the near future. Whether it is use or non-use value, option value considers the preservation, protection, or indecision about the possibility of benefiting from that good or service in the future. In essence, most of the time it is a postponement of the utilization of the respective resource in the productive process. It is similar to the insurance or future employment value of a present benefit

(continued)

Table 5.3 (continued)

The value type	Meaning and specificity of assessment
Sentimental value	It is associated with emotional dependence on a specific good with the potential for economic capitalization and makes sense especially in the case of selling or changing the intended use of that good. In general, land, forest, water gloss are valued with this kind of value, the result is rather keeping in the current state and investing only for conservation and maintenance
Safety value	It is associated with the resilience of environmental factors and natural resources, resilience referring to the ability of an ecosystem to maintain its basic functions and to protect against exploitation for productive purposes. Most of the time, it takes control in conditions of disruption or economic, human, social aggression. For example, banning deforestation in areas at high risk of flooding or banning hunting and fishing in certain seasons

Table 5.4   Solutions and intervention stages for the sustainable use of resources in the production-consumption process (SRCP, 2015)

Intervention area	Solutions	Expected results
Improving the production process	– Reducing the consumption of raw materials through more careful analysis and planning of consumption needs; – Reducing the consumption of energy, utilities and other vital resources; – Reducing the use of substances harmful to the environment; – Introducing reusable waste into the production circuit; – Encouraging the circular economy;	– The promotion, on an extended scale, of sustainable production; – Encouraging and determining balanced and limited consumption and real consumption needs; – The appearance of more and more production units following sustainable production and consumption systems; – Competitive sustainable businesses;
Associating producers with the requirements and principles of sustainability	– Optimization and efficiency of production processes; – Inserting the principles of sustainable development into the production objectives; – Improving the production capacity, through innovation and re-technology;	– Sustainable and healthy economic environment; – Economic environment reconfigured in the direction of sustainability; – Functional policies and strategies in favor of the protection, valorization and conservation of natural resources;
Market, consumers and end users	– Reduction of production costs, therefore of consumer prices; – Minimizing risks by adopting environmentally friendly procedures; – Acceptance of modern technological improvement solutions;	– Increasing the technical and economic life of the products; – Increasing the performance and quality of products intended for consumption;

successfully implemented that aim to achieve the objectives of sustainable development, the sustainable use of resources throughout the life cycle of the production-consumption network or the efficient and eco-economic capitalization of natural resources. Some examples of the many successful projects implemented in Romania are: BridgeCE (partners from Romania, Belgium, Greece and Slovenia, with aim to create opportunities to switch to circular economy systems and stimulation of green business), Green Innovation in the Fashion Industry Management (partnership with Romania, Greece, Italy, Germany, Spain for training and education for the development of new ideas for a greener textile sector), AX Smart Education for Corporate Sustainability Reporting (partners from Romania, Poland, Czech Republic, Spain, Slovenia enrolled on platforms developed to support corporate responsibility according the objectives of sustainable development), Compost Academy (collection of biowaste that can be turned into compost and collaboration with the authorities to facilitate the transition to green and sustainable cities), Green Environment (for used oil collection and recovery of waste for the prevention of ecological problems), Bonapp.eco (for the responsible consumption by collaboration with economic units from public food in order to avoid food waste), etc.

5.6   An Eco-Economic Evaluation of Natural Resources: A Case Study

For the practical demonstration of what mentioned so far regarding the perception of natural resources as inputs for production necessary for any kind of outputs for consumption, it will be presented a practical study of eco-economic evaluation of a well-defined area with natural resources. The purpose of the study is to highlight the reconsideration of the way of using natural resources and is a synthesis of an eco-economic analysis project that foregrounds the decision on how to use a territory included in the category of Natura 2000 sites (Kettunen et al., 2009). The purpose of the project for which we present the synthesis is to choose between the current variant of use of the resources on this site (BAU—Business as Usual) and the proposed version under sustainable use (SEM—Sustainable Ecosystem Management). The dilemma that determined the eco-economic evaluation of the inventoried natural resources is between the decision to conserve biodiversity and resources in that area (even if economically it seems not to be advantageous) or to continue their exploitation. The project was implemented by an interdisciplinary team, in an area attached as a component of Natura 2000 sites. Given that, currently, most guidelines in economic evaluations are focused on profitability, the project succeeded, through rigorous eco-economic estimations, to determine economic values considering the importance of natural resources in the long term. In essence, the value of the natural resources in this Natura 2000 site has been determined. The option of considering the contribution of these resources, found in the analyzed natural ecosystems, to human well-being in the short and medium term was chosen.

The site analyzed and evaluated eco-economically is called "Dobrina Forest", located in the vicinity of Huşi, Vaslui County, Romania. The objectives pursued started from the finding that the natural resources in the area chosen for presentation and analysis are not sufficiently well preserved or are not appreciated from the perspective of sustainable development requirements. The objectives were formulated in such a way as to allow a better understanding of the advantages that nature makes available to the human species and, implicitly, to the socio-economic environment. The objectives formulated in the project were:

O1. Inventory of natural resources, as ecosystem assets, from the Dobrina Forest site;

O2. Assessment of the value of ecosystem goods and services from the Dobrina Forest site;

O3. Identifying and presenting the benefits of conserving natural resources from the analyzed site;

O4. Improving the decision-making context regarding the exploitation of resources from the site.

The context in which the idea of implementing the project was started was based on the requirement of sustainable exploitation of natural resources, in general, and those belonging to the Dobrina Forest site, in particular. In comparison with the traditional economic-financial evaluation, the following are taken into account: the time factor and the rigorous inventory of available resources, as well as all the benefits they offer in the production-consumption process. The measurement of economic growth through indicators, such as GDP, total value of investments, available assets, is no longer sufficient nor relevant as long as in social terms, such as the field of human health, the level of poverty, the quality of life including mental health, and the level of education. There were some significant correlations between the indicators according to the statistical analysis (Oehler-Şincai, 2014).

Romania, in general, and Natura 2000 sites, in particular, are faced with: the lack of forest and biodiversity protection systems, the pollution increasing, the lack of investment in environmental infrastructure, climate changes, floods that alternate with droughts major, generalized degradation, etc. Among other things, the study aims to encourage investments to finance the conservation and protection of the Dobrina Forest site. The eco-economic analysis and sustainable exploitation of the resources in this site starts from the availability of ecosystem services of this site. Conceptually, ecosystem services represent the totality of benefits, tangible and intangible, that natural or natural-anthropic ecosystems provide to society. Without ecosystems and the services, they provide, human well-being would be in imminent danger. An ecosystem, that is, that natural unit that includes all living organisms that interact with each other, constitutes a higher level of organization of all the components of nature. For the analyzed area, but also in the generalized practice of ecosystem analyses, 4 large categories of ecosystem services are identified, interdependent and influencing each other. These service categories are explained in Table 5.5.

Table 5.5 Categories of ecosystem services (Miron, 2019)

Service category	Features and functionality
Procurement and supply services	Physical resources and products, usable as inputs in production. These services provided by natural ecosystems directly contribute to the production process as food, raw materials, water supply
Regulation services	Are correlated with the regulation of natural processes in ecosystems and refer to the fixation and storage of carbon, the regulation of extreme phenomena such as floods, the regulation of air, water and soil pollution
Cultural services	Almost exclusively include the non-material benefits that people derive and use from direct contact with nature. The most common such service, difficult to quantify by direct methods, is recreation, along with other pleasures that people benefit from through contact with nature and its knowledge. Equally important and still difficult to assess directly are the spiritual values provided by these services
Support and support services	It refers to those processes that support and enable the provision of other ecosystem services, such as: primary production, exploitation of raw materials and inputs for production, soil formation, nutrient cycling

Although the differentiation of these ecosystem services may induce the idea of a major difference between the economic and the ecological approach, it is demonstrated that, regardless of the approach in the assessment of ecosystem services, all together and each individually or combined contribute to social well-being and individual, supporting economic development and preserving natural heritage.

The discussion about value can become subjective and uncertain, but this is precisely the reason why, in terms of sustainability, the devaluation or undervaluation of natural resources and all that means environmental factors is not accepted. In the analysis of this study regarding the ecosystem values of the "Dobrina Forest" site, several categories of values and benefits will be considered: ecological, socio-cultural, and economic.

5.6.1   General Description of the "Dobrina Forest" Site

Dobrina Forest, located 5.1 km. by the town of Huşi, in Vaslui County in Romania, is a Site of Community Importance (SCI), being an integral part of the European Natura 2000 ecological network in Romania. It has an area of 8518 ha. The main classes of habitats identified in the site and inventoried for the eco-economic evaluation necessary for the decision to integrate into the production-consumption networks are: oak, hornbeam, beech forests, deciduous thickets (96%)–8177.28 ha., meadows and semi-natural hayfields (3.0%)–255.54 ha, arable land (0.7%)–59.63 ha, other lands (0.1%)–8.52 ha, fresh flowing waters (0.2%)–17.04 ha, etc. The purpose for which it was included as a community reference site in the Natura 2000 network is to preserve the habitats it contains, given that the pedo-climatic conditions are above-average and allow good vegetation maintenance for use in productive use systems. The economic activities carried out in the villages bordering the site are: exploitation of agricultural land, field crops, horticulture, animal husbandry, exploitation of wooded lands. The elements of tourist attraction inside and around the evaluated site offer the possibility of practicing many types of tourist activities: (i) ecological tourism: supported by landscape elements and valuable flora and fauna; (ii), unorganized weekend tourism: activity that makes it difficult to achieve the conservation objectives due to various disruptive anthropogenic activities (waste storage, soil compaction, fire lighting, etc.); (iii) cultural tourism: associated with the presence of monuments of historical value. The threats already manifested, but also potential, to the Dobrina Forest site are: cutting trees (including illegal ones), poaching and destruction of bird nests, unorganized tourism, waste storage, anthropogenic fires to clear agricultural land adjacent to the protected natural area, the grassy layer affected by unregulated grazing and tourism and the change in the phreatic regime in recent years, which led to severe premature drying.

Regarding the sustainable management of the resources in the assessed site, it is noted that there is still no management structure, nor has any specific site management plan been developed.

The eco-economic evaluation of the ecosystems in the "Dobrina Forest" site is done with the two alternative scenarios (BAU and SEM), starting from two main options:

1. The future exploitation of the site will be done under the same conditions as before, that is, the resources are valued as economic, monetary value, given their character as inputs for the production and consumption processes; this is the "Business as Usual" (BAU) scenario;
2. Future exploitation will be done in accordance with the requirements of sustainability, the resources being exploited further, but in a balanced and responsible way; this is the "Sustainable Ecosystem Management" (SEM).

The transition from the BAU scenario to the SEM scenario involves the analysis of a set of reference indicators, whose values are compared: Net Updated Value (NPV), number of jobs, unemployment rate, collected revenues, fiscal impact, production in ecological system, the value and return on investment, the value of natural capital, the level of quality of life, etc.

Following the inventory of ecosystem services in the Dobrina Forest, which are evaluated in an eco-economic regime, given the proposal to switch the use of resources from the BAU usage scenario to the SEM scenario, these services are grouped as follows, by category and content:

1. Supply and Supply Services: agricultural food production, non-agricultural biomass, animal biomass, medicinal resources, raw materials and materials (work-wood, firewood), game, water, biological diversity resources;
2. Regulatory Services: air and water quality regulation, disease and pest control, climate regulation and carbon sequestration, flood control and protection, pollination;
3. Cultural Services: tourism and recreation, ecotourism, landscape, educational values.

4. Support Services: primary production (photosynthesis), nutrient cycle, ecological interactions, evolutionary processes.

All these services have been inventoried, evaluated on several value categories and valued for both reference scenarios with detailed calculations. A summary of the assessment of resources inventoried from the "Dobrina Forest" site is presented in Table 5.6 (where "yes" indicates the validity of those value categories for the respective services and "no" indicates that those services do not have their respective value listed).

For the value analysis in monetary terms (in lei—the national currency), the respective values were brought to a common denominator, net present value (NPV), to be able to make the comparison between the two scenarios (BAU and SEM), presented in Table 5.7.

For each service category, the following conclusions are drawn regarding the decision to maintain the current BAU scenario or move to the SEM scenario. Supply Services: exploitation of raw materials is the service with the highest significance, ornamental resources are exploited and exploited minimally or not at all, the most practiced currently is wood exploitation, the preservation of the Dobrina Forest site is more effective, by preserving the current situation, i.e., reducing the volume of harvested timber and expanding the area of forests with a special protective role, the negative effects, specific to the BAU scenario, will be substantially reduced.

Table 5.8 compares the calculated values for each category of services provided by the eco-economically evaluated site.

Table 5.6 The eco-economic values of the ecosystem services provided by the Dobrina Forest

Ecosystem services	Value directly used	Value indirectly used	Optional value	Unused value
Provision services	YES	NO	YES	NO
Regulation services	NO	YES	YES	NO
Cultural services	NO	YES	YES	YES
Support services	YES	YES	YES	YES

Table 5.7 Net present value (NPV) of ecosystem services in the SEM version and the BAU version (lei) of the resources of the Dobrina forest

The ecosystem service	VAN in BAU version	VAN in SEM version	VAN difference between SEM and BAU
VAN supply services	174,928,07	191,847,9	16,919,82
NAV adjustment services	98,395,19	100,993,86	2598,67
VAN cultural services	1749	2169,92	420,92
Total NPV	275.072,26	29.501,68	19,939,41

Table 5.8 Net present value (NPV, in USD) of SEM versus BAU

The ecosystem service	SEM value	BAU value
Food in agriculture	33,88	130,06
Non-agricultural biomass	883,15	870,64
Biomass for animals	188,02	185,60
Raw materials/materials	931,58	1006,91
Biomass fuel	253,98	274,33
Medicinal resources	112,58	110,27
Ornamental resources	0,00	0,00
Resources for bioprospecting	0,00	0,00
Water supply	5,44	5,44
Total supply services	2408,63	2583,25
Total NPV	41,820,62	38,132,29

According to the BAU scenario, it is assumed that timber hasting will continue to grow slightly to provide raw material to productive sectors based on wood processing and firewood or work-wood for local communities. The SEM scenario starts from the following premises: expansion of the areas with a special role of protection and conservation; sustainable exploitation of non-timber and game products of the forest; granting compensatory payments for private forest owners. For the cultural services: cultural heritage has limited significance for the site, but with the potential of capitalization in the future; ecological education little to no capitalization at present; spiritual and religious goods: limited significance for the site, but with the potential to be exploited in the future; goods for health, tourism, and recreation: existing, of high value, but not capitalized. For the regulatory services: the highest contribution to the total value is made by the water purification and waste management services (67%), the water leakage regulation service (6.94%), and the services for maintaining the genetic diversity of species (10.72%); under these conditions, it is more effective to preserve this site.

The major difficulty in eco-economic assessment was not in inventorying ecosystem services, but in quantifying them and assigning a monetary value to be validated in economic systems. A very important principle is that of the correct classification of ecosystem services, from the point of view of the purpose of their use, given that some are intermediate and others are final. Major difficulties arise when considering territories that can be used for both agricultural and non-agricultural production. Another important principle is that of the exhaustive evaluation of ecosystem services that must consider and calculate the interdependencies between the different services or goods and the possibility of their successive transformation, up to the final stage of direct capitalization, of the type of economic production process. The intermediate components, such as air and water quality, medicinal resources, the touristic landscape, are evaluated somewhat more easily, through indirect methods. Others, however, are very difficult to quantify, including with such methods; examples: the silence, the stressful natural setting, the beauty of the water, the clouds, the admiration of the fauna, etc.

The real problem, and to a great extent difficult, is given by the analysis of how to use an ecosystem in all its complexity. A vital element is lost sight of in this context: the potential for future generations to enjoy the same benefits is limited. Furthermore, as ecosystems and embedded resources end up being severely damaged, restoration becomes much more expensive and time-consuming and, in some cases, even impossible.

Well-being is a multidimensional concept that includes physical, material, social, mental well-being and the general level of satisfaction; well-being is that ideal state to which both society and the individual aspire, through the way of producing, saving and consuming. (Felce & Perry, 1995). The functionality of these links is valid locally, regionally, nationally and internationally, short-term and long-term. Between ecosystem services and human well-being there is a relationship of continuity and dependence on a circuit involving numerous influencing factors. Ecosystems, through an accumulation of created structures and functional processes engaged, provide varied and necessary ecosystem services for the production-consumption chain. Through their functionality as inputs in production processes, ecosystem goods and services are transformed into (quantifiable) monetary values and benefits that will later be included in the gearing of decision-making systems in several fields: economic, administrative, legislative, governmental, consumer market, international relations, etc. First of all, however, the administrative systems, responsible for the management and administration of monetary values and the benefits provided by ecosystems, they will assume the decisions regarding how to manage these values. It is demonstrable that the responsibility assumed by the entities in the general administrative apparatus at the national level generates, directly and indirectly, various forms of human well-being. In this context, for the well-being—which can be economic, social or related to ecosystem protection—there are measurable indicators: sustainable development indicators, Human Development Index, Happy Planet Index, etc.

5.7 Conclusion

Finally, is noticed that the concern for a sustainable management of natural resources can be extended and applied to multiple levels. Along with production and consumption, as basic subjects in the analyzes and reports of representative institutions, the concept of sustainable development is becoming more and more pragmatic. Therefore, a direct link of mutual influence between the issue of natural resources and that of sustainable development it was highlighted.

The paper managed to show the eco-economic evaluation method of a set of resources from a protected area, an aspect that indicates the possibility of expanding the analysis at the national level. It is possible to learn that a healthy relationship between governments, European institutions and local communities can be developed, conditioned by the reconsideration of the development of humanity objectives.

Consumption and production, basic elements of the economic environment, can be adjusted to the new requirements for stimulating and revitalizing natural resources. So, without a primary consideration of the value of natural resources, both production and consumption, and the quality of life, too, can be affected on the long term.

References

Angelova, M. Dimitrova, T., & Pastarmadzhieva, D. (2021). The Effects of globalization: Hyper consumption and environmental consumer behavior during the Covid-19 Pandemic. *International Journal of Economics and Business Administration, IX*(4), 41–54. https://doi.org/10.35808/ijeba/733

Anonymous. (2023). *Sustainable development department. national strategy for the Romania's sustainable development–2030. Romanian Government official website.* Retrieved from https://dezvoltaredurabila.gov.ro/ (Access date: 14 July 2024).

de Jesus, A., Lammi, M., Domenech, T., Vanhuyse, F., & Mendonça, S. (2021). Eco-Innovation diversity in a circular economy: Towards circular innovation studies. *Sustainability, 13*(19), 10974.

EC. (2019). *The European Green Deal. European Commision.* Retrieved from https://commission.europa.eu/strategy-and-policy/priorities-2019-2024/story-von-der-leyen-commission/european-green-deal_en (Access date: 18 July 2024).

ERDF. (2024). *European Regional Development Fund.* Retrieved from https://ec.europa.eu/regional_policy/2021-2027_en (Access date: 23 July 2024).

Felce, D., & Perry, J. (1995). Quality of life: Its definition and measurement. *Research in Developmental Disabilities, 16*(1), 51–74. https://doi.org/10.1016/0891-4222(94)00028-8

Herrmann-Pillath, C. (2013). Foundations of economic evolution. In *A Treatise on the Natural Philosophy of Economics (New Horizons in Institutional and Evolutionary Economics series).* Edward Elgar Publishing Ltd.

ISO. (2015). *International Organization for Organization. Standards: 14000 family (Environmental management), 20400 family (Sustainable procurement), 26000 family (Social responsibility and sustainable development).* Retrieved from https://www.iso.org/standards.html (Access date: 12 July 2024).

ISO. (2024). *International Organization for Organization. Standards: 14000 family (Environmental management), 20400 family (Sustainable procurement), 26000 family (Social responsibility and sustainable development).* Retrieved from https://www.iso.org/standards.html (Access date: 22 July 2024).

Kettunen, M., Bassi, S., Gantioler, S., & Ten Brink, P. (2009). Assessing socio-economic benefits of Natura 2000—A Toolkit for Practitioners. Output of the European Commission project Financing Natura 2000: Cost estimate and benefits of Natura 2000. *Institute for European Environmental Policy* (IEEP).

McCarthy, D., & Morling, P. (2014). A Guidance Manual for Assessing Ecosystem Services at Natura 2000 Sites. Produced as part of the Natura People project, part-financed by the European Regional Development Fund (ERDF) Programme 2007–2013. *Royal Society for the Protection of Birds*: Sandy, Bedfordshire.

Michieli, I. (1993). *Appraisal treaty.* Edagricole.

Miron, V. (2019). *The economic value of biodiversity and ecosystemic services, in the framework of Project Integrating the biodiversity conservation priorities in the policies of territorial planning* and practices for using the lands from Moldavia; granted by Global Environment Fund (GEF) and implemented by UNEP (United Nations for the Environment Protection).

NIS. (2018). *The indicators of sustainable development for Romania. National Institute of Statistics in Romania.* Retrieved from https://insse.ro/cms/files/Web_IDD_BD_ro/index.htm (Access date: 24 June 2024).

NIS. (2023). *National indicators for sustainable development, Horizon 2030.* Sustainable Romania: National Institute of Statistics. Publishing House of the National Institute of Statistics, Bucharest

NIS. (2024). *Statistical indicators of sustainable development.* National Institute of Statistics in Romania. Retrieved from https://insse.ro/cms/files/Web_IDD_BD_ro/index.htm (Access date: 26 July 2024).

Oehler-Şincai, I.M. (2014). Searching a veritable indicator of the wellbeing. In *International conference „toward a good society. European perspectives"*, Bucharest (October 2013); published in Quality of Life, XXV, No. 1.

Rajović, G., & Bulatović, J. (2017). Natural resources, classification of natural potential, sustainable development. *World News of Natural Sciences, 6,* 12–27; EISSN 2543–5426.

SRCP. (2015). *Adaptation after National Center for the Sustainable Production and Consumption, Eco-innovation—A way to the sustainable production and consumption, Project co-funded by the Swiss-Romanian Cooperation Programme.*

Strunz, S. (2014). The German energy transition as a regime shift Author links open overlay panel. *Ecological Economics, 100,* 150–158. https://doi.org/10.1016/j.ecolecon.2014.01.019

UN. (2023). *Transforming our world: the 2030 Agenda for Sustainable Development. Water Conference 22–24 March 2023, New York.* Retrieved from https://sdgs.un.org/2030agenda (Access date:15 May 2024).

Chapter 6
Activities on the Efficient Use of Water in Agriculture in Türkiye Under Climate Change

Ahmet Şeren® **and Mehmet Uğur Yildirim**®

Abstract It has been realized once again why water is so important during the epidemic process, with the droughts experienced in recent years associated with the decrease in precipitation and increase in average temperature as a result of global warming. At this point, food security and supply based on irrigated agriculture have become the most important issue for humanity. The limited crop pattern in rainfed agricultural areas changes in irrigated areas. Agricultural irrigation is still the largest water using sector in the world and in Türkiye. About 69% of the world's water resources are used in agriculture, 19% in industry, and 12% for domestic purposes. In Türkiye, these ratios are 76% in agriculture, 24% in industry and for domestic purposes, respectively. The limited water resources and the increasing demand in all sectors in Türkiye make it necessary to use the existing water resources in the most efficiently ways. Therefore, in water management, it is necessary to ensure a fair sharing and effective use of water between sectors in harmony with the environment. This chapter deals with the works under the structural and nonstructural measures implemented in Türkiye for effective use of water in agriculture.

Keywords Irrigated agriculture · Irrigation efficiency · Modernization · Water saving

6.1 Introduction

Water is an indispensable input for all production processes, from agriculture to industry, from energy to the service sector. Reasons such as the current growth rate and water consumption habits exert significant stress on water resources. However, the

A. Şeren (✉) · M. U. Yildirim
State Hydraulic Works (DSI), Ankara, Türkiye
e-mail: aseren@dsi.gov.tr

M. U. Yildirim
e-mail: muyildirim@dsi.gov.tr

adverse effect of climate change on the hydrological cycle and spatial and temporal distribution of water is increasing year by year.

Climate change predictions show that the Mediterranean Basin, including Türkiye, will be seriously affected by the increase in temperature and decrease in precipitation. It is estimated that this situation will increase water stress and cause droughts to occur more frequently and severely, resulting in water shortages, increases in forest fires, loss of biodiversity, and loss of income in agriculture and tourism. According to the data of the United Nations (UN), while global water stress, which is expressed as the ratio of water withdrawn for agriculture, drinking and industry purposes to available water, was at a manageable level with 18.2% in 2020, 2.4 billion people have been living in the region exposing to extreme water stress in some cases since 2022 (UN, 2023). As shown in the infographic above (Fig. 6.1), based on World Resources Institute (WRI) projections of temperature increment of between 2.8 and 4.6 °C by 2100, 51 of the 164 countries and territories analyzed—31% of the world's population—are expected to be faced with high to extremely high water stress by 2050 (Armstrong, 2024).

In the Aqueduct 4.0: Updated Decision Sensor Global Water Risk Indicators study prepared by World Resources Institute experts determined that the world is facing with an unprecedented water crisis, due to the fact that 25 countries making up a quarter of the world's population are currently exposed to extremely high water stress every year, globally, and approximately 4 billion people, half of the world's population, are exposed to water stress for at least one month a year, that this rate could be close to 60% by 2050. Additionally, $70 trillion of gross domestic product (31% of global Gross Domestic Product, GDP) will be subject to high water stress, compared to $15 trillion (24% of global GDP) in 2010. They found that only four

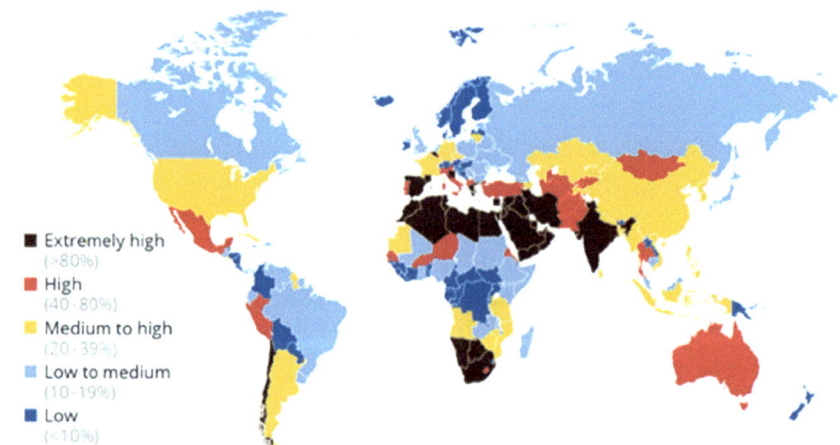

Fig. 6.1 Geographical regions in the world where water stress will be the highest by 2050 (Armstrong, 2024)

country—India, Mexico, Egypt, and Türkiye—accounted for more than half of the GDP exposure in 2050 (Kuzma et al., 2023).

On the other hand, food security and supply based on irrigated agriculture have become the most important issue for humanity. The limited crop pattern in rainfed farming areas varies with the developed irrigation projects. The increase in production value and gross national agricultural income in irrigated areas in Türkiye is at least 5 times compared to rainfed agricultural areas. Considering data in the world, agricultural added value increased by 84% between 2000 and 2021, reaching 3.7 trillion USD in 2021 (FAO, 2023). On one hand, this situation serves the purpose of reducing poverty, which is one of the goals of rural development; on the other hand, it prevents migration because it enables to improve the standard of living.

Agricultural irrigation is still the largest water using sector both in the world and in Türkiye. 69% of the world's water resources are used in agriculture, 19% in industry, and 12% for domestic purposes. However, distributions of sectoral water uses may vary depending on the development levels of countries. Agricultural water use largely depends on both the climate conditions and the place of agriculture in the economy. In South Asia, agricultural, drinking, and industrial use rates are 91%, 7%, and 2%, respectively, while in Western Europe these rates are 5%, 23%, and 73%, respectively, (FAO, 2024).

The countries having the highest shares in agricultural use are mostly in Africa and Asia. Somalia's share is over 99%. One of the common characteristics of these countries is income level. The World Bank classifies 6 of the top 20 countries as low-income, 12 as lower-middle-income, 1 as upper-middle-income, and 1 as high-income (FAO, 2023).

According to the above data and information, irrigated agriculture is not only very important for food security but also the sector that uses the more water. Therefore, it is necessary to use water efficiently and economically in existing irrigated agriculture and to take some precautions to save water and increase water productivity in agricultural irrigation for sustainable agriculture and food security. In this context and in this chapter, the works of the State Hydraulic Works (DSI), which is the only institution responsible for the management of water resources and the establishment of irrigation systems and infrastructure in Turkiye, is presented.

6.2 Climate and Water Resources in Türkiye

Türkiye is located in the semi-arid climate zone of Mediterrenian Region, and precipitation varies greatly depending on location and time. The long-term average anual rainfall is 574 mm. The Eastern Black Sea Region receives the most rainfall in the range of 1200–2500 mm yr^{-1}, and the Central Anatolia Region receives the least rainfall with an average of 250–300 mm yr^{-1}. Although the water potential is approximately 450 billion m^3 according to the annual average precipitation, within the framework of today's technical and economic conditions, the average amount that can be used annually for various purposes is 112 billion m^3. Of this, 94 billion

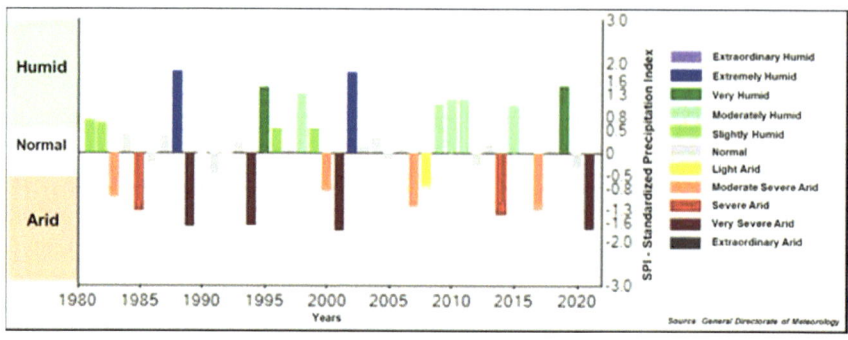

Fig. 6.2 Drought analysis in Turkiye (TRGM, 2022)

m^3 is surface water and 18 billion m^3 is groundwater potentially. Thus, Türkiye is not a water rich country, and the distribution of available water resources throughout the country is not equal. While 28% of the total population lives in the Marmara Region, the basins here collect only 4% of the total flow. Due to population growth, the annual amount of usable water per person was 1652 m^3 in 2000, 1544 m^3 in 2009, 1346 m^3 in 2020, and 1313 m^3 in 2023. It is expected that population in Türkiye will reach 100 million in 2050 and the amount of water per capita will decrease to 1120 m^3. If these expectations come true, Türkiye will be in the position of countries suffering from water stress (DSI, 2024a).

According to the drought analysis conducted by Turkish State Meteorological Service (MGM) in Türkiye between 1981 and 2021; 41-year period, 11 years were classified as dry, 13 years as wet, and 17 years as normal. The 2000–2001 water year is the driest, followed by the 2020–2021, 1988–1989, and 1994–1995 water years, respectively (Fig. 6.2). Although no drought is observed in the analyzes made based on average precipitation, drought occurs almost every year in some basins.

Increasing drought severity accelerates the depletion of groundwater in various basins, especially in the Konya basin, it threatens the long-term sustainability of agriculture and causes the formation of large sinkholes. On the other hand, during the varying climate conditions in recent years, especially the precipitation regime and amount have changed. The, precipitation that does not fall for a very long time has decreased to the total amount of a few months in a much shorter period of time. The temperatures in winter and sping months have increased. The period suitable for agriculture has been extended in almost all basins, and the amount of snow that has fallen decreased and remained in a shorter time duration on the soil. As a result of its decrease, the amount of water used in agriculture increases on an annual basis.

It is of great importance to store the water source, which has no alternative for all living things, for use in dry periods, especially in countries with irregular rainfall regimes like Türkiye. In geographical regions of Türkiye, except the Eastern Black Sea coastline, irrigation becomes inevitable because the total rainfall during the crop growth period does not meet the total crop water consumption. This requires to construct storage facilities, such as dams, ponds, for controlling water resources.

In Türkiye, 43.9 hm^3 (76.7%) of the 57.3 hm^3 water consumed annually is used in agricultural sector and the remaining part is used in the service and industrial sectors. In 2050, the total water need is expected to exceed 65.00 hm^3, 52.7 hm^3 of which will be the need for agricultural water consumption (DSI, 2020).

From this perspective, steps toward using water economically in this area are very important. The reality of having limited availability of water resources and the increasing demand in all sectors in Türkiye make it necessary to use the existing water resources in optimum efficiency.

6.3 Activities Including Structural Measures for the Efficient Use of Water

6.3.1 Conversion of Existing Open Systems to Pressurized Piping (Closed) Irrigation Systems

The pressurized irrigation systems and modern irrigation methods are considered as applications that allow the most effective use of water and provide the highest measurable or unmeasurable economic and social benefits in the long term, providing that necessary technical, economic, and social conditions are met. Use of modern irrigation systems are inevitable for water saving, therefore, it should be understood that farmers are key factors of reaching the goals. Farmers' intent to and acceptance of modernization may vary from one region to another in Türkiye. Modernization policy is realized in accordance with this reality. Modernization of irrigation systems has resulted in a significant increase in both water savings and farmers' net income (Çetin et al., 2024).

Until 2003, irrigation networks were mainly classical open canal system, canalette system and low pressurized system. However, new irrigation projects and also renovation works were carried out for the modernization of irrigation infrastructures and systems in operation in recent years. This implementations ensure minimum land loss, high conveyance efficiency, easiness of placing water metering facilities for volume-based operation. The high-pressurized network system is preferred due to the possibility of applying a water usage service fee, easiness of operation, long facility life, lower maintenance, and repair costs compared to other systems, not being able to intervene much in the system by the farmers and saving water.

The rate of closed irrigation networks, which was 6% before 2003, has now reached 35% (Fig. 6.3). The rate of pressurized piped irrigation networks is expected to reach 45–50% in the near future.

On the other hand, old open canal irrigation systems, which were built and put into operation in the past years, are modernized within the scope of renovation projects. In order to ensure that irrigation systems fulfill their functions, continue

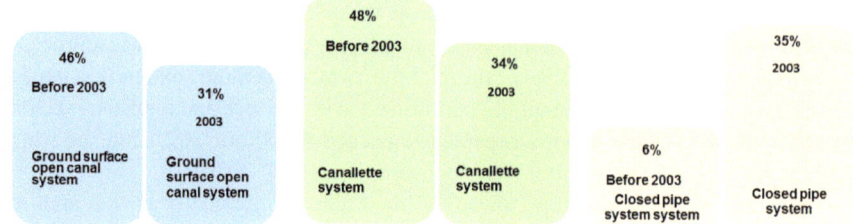

Fig. 6.3 Conversion in irrigation water conveyance systems in Turkiye

their contribution to the national economy, improve the benefit period and conditions of farmers, ensure water saving through the continuity of operation and maintenance activities, the needs of the facility that cannot be met by maintenance and repair works are met within the scope of the "Modernization Project". Within the scope of the Modernization Projects of the irrigation systems in operation, planning, projecting, and construction works are continuing in an area of approximately 1.3 million hectares (Fig. 6.4).

On the other hand, World Bank loans are used to finance some of the renovation works. Eight irrigation networks serving a total area of 111,000 ha, four of which serve an area of approximately 50,000 ha within the scope of the Türkiye Irrigation "Modernization Project" and four serving an area of approximately 61,000 ha within the scope of the Türkiye Water Circularity and Efficiency Improvement Project, will be converted to a pressure piped network.

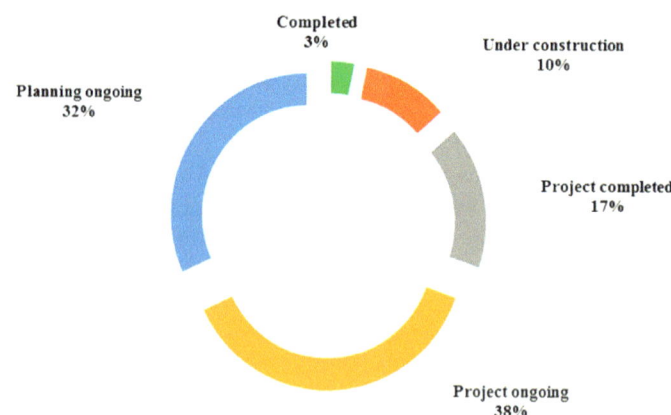

Fig. 6.4 Distribution of modernization projects in Turkiye

6.3.2 Switching to Prepaid Water-Meters

An effective irrigation management can be achieved by measuring the water used at all levels, completing in-field development services and complying with water distribution programs. In this context, activities are intensively continuing to convert water measurement facilities in irrigation systems that have been put into operation into electronic systems that will allow central monitoring and evaluation.

Water-meters are used to ensure efficient use of irrigation water through accurate measurement, to provide the water the plant needs effectively and economically, and to ensure equitable distribution of water in irrigation systems. In pressurized pipe irrigation systems where the irrigation infrastructure is suitable, water users have been encouraged to use less water by switching to prepaid meter applications for the purpose of measured and controlled use of water, enabling volume-based charging of water price instead of area-based.

According to the monitoring and evaluation results of irrigation systems, the irrigation efficiency in piped irrigation networks, where the water usage service fee is determined in cubic meters, was 84% and the water use per hectare was 5092 m^3 (Fig. 6.5). Therefore, 48% savings were achieved in irrigation water use compared to the country average of 9741 m^3 ha^{-1} (DSI, 2024b).

Considering the results of monitoring and evaluation criteria, there is an increase in economical irrigation methods due to factors such as improvement in the irrigation infrastructure, the increase in closed irrigation systems, the increasing importance of effective use of water due to droughts, and the awareness of water users.

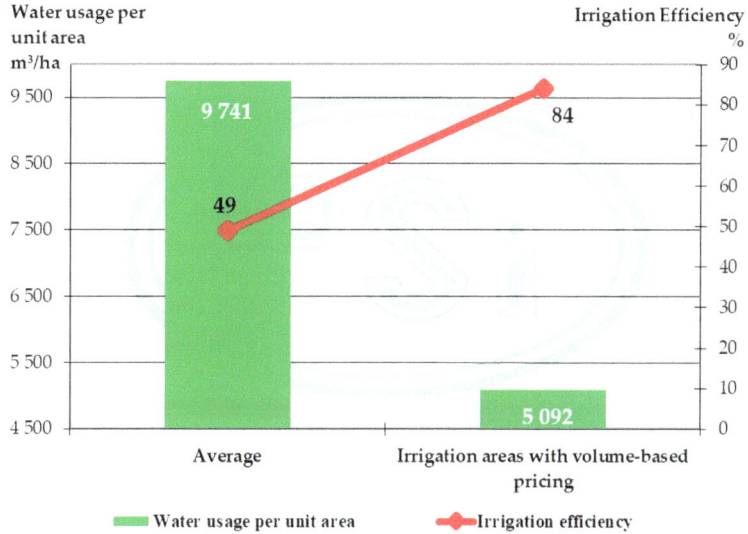

Fig. 6.5 Difference between country average water use and pricing based on volumetric measurement (DSI, 2024b)

In 2003, when DSI decided a policy change regarding the transformation from open irrigation systems to pressurized pipe irrigation systems, the area irrigated by drip irrigation system was 1% and the area irrigated by sprinkler irrigation system was 7%, and the area irrigated by surface irrigation methods was 92%. In 2023, all these are 17%, 24%, and 59%, respectively. On the other hand, in areas where volume based water use service fee tariff is applied, the total rate of sprinkler and drip irrigation method is 74%.

Therefore, activities are continuing to implement a water use service fee tariff on a m^3 basis in all piped irrigation areas.

Although there is an increase in the application of economical irrigation methods in the field due to reasons such as the increase in droughts due to the effect of global climate changes, the widespread use of pressurized pipe irrigation systems that reduce transmission and evaporation losses, and the awareness of water users, increasing agricultural supports for the use of irrigation methods suitable for the irrigation system in irrigation areas with suitable infrastructure encourages the use of methods.

6.3.3   Automation in Irrigation System

The main task of irrigation management is to meet the irrigation water demands within the irrigation area at the desired time and amount, within the existing water potential, the capacity of the irrigation system and the actual needs, by ensuring fair sharing and flow security from upstream to downstream (Anonymous, 2017).

Some findings suggest that reductions in water use by irrigated agriculture will not come from the technology itself. Rather, measures like limiting water allocation will be needed to ensure a sustainable level of water use (Perry et al., 2017). The automation and decision support systems in irrigation management ensures not only the conveyance of water from its source to the field, but also the control of distribution within the field and the limiting water allocation in optimum way.

In automated systems, soil, plant, and atmosphere sensors placed in the irrigation field transmit instantaneous values to the data processing center for full-time irrigation management, where the water need in the field can be determined on a general and tertiary basis automatically by the relevant software. Thus, water intake into the network is ensured in proportion to the actual needs. At the same time, water savings can be achieved by using automation systems that can instantly change the water distribution within the network and the crops can be watered on time without getting stressed (Anonymous, 2017).

In order to prevent excessive water use and rationalize the management of irrigation networks, automatic decision support systems have been developed where users plan irrigation time by using user-friendly, state-of-the-art software, and applying remote sensing techniques through the geographic information systems (GIS) models and satellite images (Natalizio, 2015).

Automation and decision support systems have begun to be disseminated for a more controlled irrigation management. External intervention can be prevented and water can be supplied to the network according to actual needs, especially with automation applications in closed irrigation systems,

The first irrigation automation projects were implemented in Muğla Bayırköy, Denizli Hasanbeyler, Aydin Plain, Isparta Seyitler, and Selevir Irrigation Schemes in Türkiye. By putting a remote controlled meter system to the hydrants, the irrigation time and the amount of water to be used for irrigation are determined by the water user organization according to the crop pattern. The water user can start irrigation through the center or via a mobile phone application without going to his land. Within the scope of Adana İmamoğlu Irrigation Smart Grid Management System Project; hydrants with fully automatic wireless communication have been installed (Fig. 6.6). Each water user can irrigate remotely, based on central control and management, through the index and valve opening rights associated with their subscription. By making the agricultural irrigation network smarter, auto-controlled irrigation is activated according to different criteria that can be determined through the central system.

With the GIS-based interface, online declaration independent of location and time, instant plant inspection tracking and management from planting to harvest, and automatic surveying can be made. A hierarchical management panel and separate management panels have been planned for the Regional Directorate of DSI, Water User Association (WUA), Technicians and Farmers. With the irrigation panel, allocation of the water distribution according to the total area and crop pattern, automatic revision of irrigation planning according to the automatic data of many integrated systems such as plant pattern, soil moisture, water pressure, current weather condition, network status, automatic revision of the irrigation planning with the dynamic cadastral structure of the declaration received according to the physical field can be accrued.

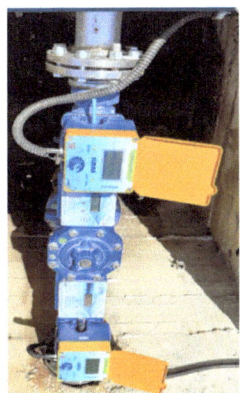

Fig. 6.6 Automation project field equipments in İmamoğlu Irrigation Scheme

On the other hand, in many irrigation systems with open channel networks, especially Aydin Söke Plain Irrigation, cover systems in backup and tertiary channels are controlled remotely.

6.3.4 Establishment of Flow Measurement Devices

In order to provide the economic and social benefits expected from the irrigation systems and to manage all according to modern operating conditions, it is necessary to measure the flows and level changes coming to the storage facilities and the amount of water used in the irrigation network, supplied from the bottom weir for drinking water, irrigation, industry, and similar purposes. Measurements are of great importance in terms of providing data for water distribution program activities to be carried out for the purpose of effective use of water and saving water, being the main data in the evaluation of demands for the rehabilitation of facilities and guiding integrated basin management activities.

It is certain that taking water into the network according to real needs and ensuring that it is used in the field according to real needs is related to the establishment of an appropriate operating system, as well as the measurement of water at every stage. For this reason, centrally monitored measurement facilities are installed in the storage facilities and irrigation networks in the enterprise to ensure instantaneous monitoring of the amount of water. The data of the installed electronic measuring devices are monitored through the established system.

6.3.5 Increasing Conveyance Efficiency by Maintenance and Renewal Projects

The irrigation infrastructures and irrigation systems used are worn out or damaged over time and with the effect of use. If timely protective measures are not taken against the effects of nature, it is destroyed and over time it completely loses its function and becomes unusable. Achieving the expected benefits from systems built for specific purposes is possible by operating the irrigation schemes in accordance with their intended purpose and periodic maintenance before malfunctions occur, and repairing them after malfunctions occur, according to the project criteria (Şeren & Çarbaş, 2015).

The operation, maintenance and management responsibility of irrigation schemes are transferred to water user organizations (WUOs) and the maintenance and repair works of the systems are carried out by these organizations. However, the irrigation performance of the systems is negatively affected due to reasons such as natural conditions, human interventions and misuse, and inadequate and timely maintenance and repairs. Considering this situation, Maintenance and Renewal Projects are implemented based on the principle of meeting maintenance and repair needs together with WUOs in order to ensure the sustainability of irrigation facilities and to improve the benefit period and conditions of farmers.

There are alternatives to reimburse WUOs after all the work to be carried out within the scope of the projects is carried out by DSI, or to reimburse the material cost after the material is covered by DSI and the labor is carried out by the water user organizations. Therefore, financial support is provided by the state to water user organizations for the maintenance and repair of irrigation facilities. Within the scope of the projects, many works such as renewal of channel panels, pump station repair, and mechanical equipment repairs are carried out.

6.3.6 Reuse of Wastewater in Irrigation

Due to the increasing population and developing industry, the need for water is increasing rapidly and natural water resources are rapidly depleting. Therefore, it is of great importance to ensure the recycling of wastewater by managing it well. Many countries that are aware of this situation use wastewater treatment technologies to protect existing water resources and quality and to enable the reuse of the resulting wastewater (Adali & Yalili, 2020).

Improving awareness level of public and using waste water in irrigation after advanced treatment are both sustainable approaches and compulsory for solution. However, the adoption and rules, preparing guidelines for using of treated wastewater in irrigation must be certainly taken into account. The impact on farmers' income of waste water in irrigation, farmers' and consumers' approaches to the subject and awareness level of them, probable effect on public health of waste water irrigation must be determined in every wastewater-based Project (Yildirim & Gül, 2018).

Taking into account the reuse of global water after advanced (tertiary) treatment in the world, 32% of recycled wastewater is used for agricultural irrigation, 20% for landscape irrigation, 19% for industrial, 8% for environmental purposes, 6.4% for recreational purposes, and 14.6% for other purposes. The use of untreated or diluted wastewater for irrigation purposes has been around for many years. The main challenge in using wastewater for irrigation is the transition from informal, unplanned use of untreated or partially treated wastewater to planned safe use (WWAP, 2017).

In the Evaluation of Used Water Reuse Alternatives Project carried out by the General Directorate of Water Management in Türkiye, alternatives for the reuse of water were examined and a total potential of 7.2 billion m^3 yr^{-1} of treated used water was determined. It has been determined that 44% of this water used in homes, industry, energy production, and agricultural sectors can be reused. Of the 3.2 billion m^3 yr^{-1} water that can be reused, 65% (2 billion m^3 yr^{-1}) is used for agricultural irrigation, 22% for environmental use, 10% for industry, 2% for groundwater recharge, and 1% for groundwater recharge. It has been evaluated that 0.1% of it can be used in landscape irrigation and 0.1% can be used in feeding drinking water resources.

The first pilot project for the use of treated wastewater in irrigation is Afyonkarahisar Domestic Wastewater Treatment Plant Irrigation. Afyonkarahisar Advanced Biological Wastewater Treatment Plant (WWTP) has a capacity of 44,000 $m^3 d^{-1}$, and the water taken from here is passed through the disinfection unit and pumped to the regulation pool built in the Ortakir Hill location in the south-east of the treatment plant, and is given by gravity to the irrigation network serving a gross area of 905 hectares. In order to meet the electricity consumption of the disinfection units and the pump station, a solar power plant with a photovoltaic panel capacity of 1.6 MW was built within the central wastewater treatment facility. The facility, whose irrigation network construction has been completed, will be put into operation after the evaluations regarding the suitability of the purified water for public health are completed.

In addition, activities on the use of WWTP waters for irrigation in the provinces of Kilis, Malatya, Ankara, Yalova, and Tekirdağ continue.

In order to encourage the reuse of wastewater, the Ministry of Environment, Urbanization and Climate Change provides up to 50% reimbursement for electricity bills to treatment facilities. This support can be increased up to 100% for organizations that reuse wastewater treated with advanced treatment techniques.

6.4 Activities Including Non-Structural Measures

6.4.1 Implementation of Planned Water Distribution

The most basic element of irrigation management activities is receiving irrigation demands. In order for Planned Water Distribution Practices to be successful, which are based on basic principles such as fair sharing of available water according to real needs during the irrigation season, effective use of water, prevention of unscheduled unauthorized illegal water purchases, and ensuring flow security in water conveyance and distribution, water users must choose the type of crops to irrigate at the beginning of the irrigation season. They are required to submit irrigation demand (irrigator information form) to the WUA, including the information and size of the parcel to be irrigated, the channel/pipeline from which the parcel will receive water, sockets, and hydrants. Then, for each water request, it is required to fill out the water request card separately.

The most important document of the General Irrigation Planning is the prediction of the irrigation water needs of these crops according to the crop water consumption values of the crop types and areas that can be irrigated. At this stage, the updated multi-year average values of crop irrigation water need (crop water need-effective rainfall) according to the region where irrigation is located and changing climatic conditions during the irrigation season, the irrigation method applied on a farm basis and the farm efficiency according to soil characteristics and conveyance efficiency parameters according to the irrigation system are taken into account. Irrigation water requirement is calculated for each irrigation scheme separately, based on the crop pattern and area expected to be irrigated. The planning is completed by determining how much of the irrigation water need can be met by comparing these calculated/ estimated values with the existing water potential in storage facilities and the expected flow values in rivers. In cases where the annual water need is not sufficient, limited irrigation practices are applied.

6.4.2   *Monitoring of Operation and Maintenance Activities by Spatial Information System (SUTEM) for Irrigation Systems*

Remote Sensing (RS) is a technology that has been widely used in agricultural production, especially in developed countries, for a long time, as well as in many areas of life, such as irrigation scheduling, water stress, pesticide application, yield estimation, land value determination, and similar processes. However, the effective use of RS depends on the use of geographical information systems (GIS) infrastructure in large areas. Depending on the developments in RS technologies, data at different spatial, spectral and temporal resolutions can be obtained in spatial analysis. These data provide the opportunity to reach accurate results for many branches of science that need spatial information. One of this spatial information is the detection, development and change of land use and vegetation, where remote sensing technology is intensively used. Since the reflection properties of each plant may vary depending on the water and chlorophyll content and cell structure of the plant, it is possible to determine the plant type, development process, productivity and yield, determine soil moisture and type, and determine irrigated and/or dry farming areas with remote sensing. (Şener et al., 2020).

Irrigated areas under water stress within the irrigation network can be identified by utilizing innovative methods based on the use of multi-spectral satellite images with different spatial and temporal resolutions. Classification of satellite images and feature extraction are among the methods used to obtain information in remote sensing. The reliability of the information to be produced is related to the classification accuracy.

DSI has implemented the Irrigation Facilities Spatial Information System (SUTEM) software to carry out irrigation management in this direction in the irrigation projects it has put into operation. The program aims to improve irrigation service delivery and ensure traceability and effective use of water resources. Important work items such as irrigation demands, general irrigation planning, water distribution programs, measurement of irrigated area, planned water distribution report, accrual and collection in irrigation facilities are carried out in a standard manner through an information technology (IT) infrastructure that will provide service from a single point to all organizations performing operation, maintenance and management activities.

There is an integration with these services in the application: TAKBİS (Land Registry and Cadastre Information System) and MEGSİS (Spatial Real Estate System) created by the General Directorate of Land Registry and Cadastre, MERNİS (Central Population Administration System) created by the General Directorate of Population and Citizenship Affairs, ÇKS (Farmer Registration System) created by the General Directorate of Agricultural Reform.

Time and economic savings are achieved by integrating land registry information verbally and numerically, introducing real persons to the system, comparing measurements of irrigated area data and subsidy applications of water users into the services of relevant administrations.

A transparent and auditable irrigation management is ensured by not accepting contradictory data by the software, determining basic data such as base fee tariffs to be applied by DSI, and tracking the data of water using institutions from anywhere.

SUTEM is designed according to the integrated working principle of 16 main modules; Registry and Identification, Facility and Inventory, Budget and Accounting, Accrual and Collection, Irrigation Management, Surveying and Field Operations, Maintenance and Repair, Water Users, Electronic Document Management System, Communication, Audit, Purchasing, Investment Expropriation, Crop Counting, Monitoring and Evaluation.

The software, which serves at http://sutem.dsi.gov.tr/, has multi-user management. A person has one or more roles; water users, water user organizations, regional and central government users. However, every user has the right to act and access information within the authority given to him by the system administrator. In this context, Water Users can make declarations, track their measurements, view their temporary and final accruals, and make requests and complaints through the system (Şeren, 2023).

6.4.3 Flow Estimation and Basin Optimization Model (ATHOM) for Dam Operation

The basis of system is to compare the available water with the water needed through an operational hydrology study in all storage structures in operation, and to plan the use of water by sectoral sharing according to the results. It is of great importance to ensure the use of existing water in irrigation areas within the program. For this purpose, operation programs are prepared for all dams in operation, updated when necessary, and operation is ensured according to this operation program. Water storage structures planned according to basin characteristics constitute the most important link of the water cycle as the insurance of the entire system and play a vital role in periods of drought and flood.

Storage structures, which produce a wide range of economic, social, and environmental benefits, from drinking water supply to agricultural irrigation, from hydroelectric energy production to flood control, enable production activities to continue without interruption, especially during periods of drought. Another point as important as the accumulation of water in storage facilities for dry periods is the timely and accurate prediction of the flows that will fill the storage facilities with water.

Knowing the flows that will come to the dam in advance is of great importance in terms of reducing flood risk, producing hydroelectric energy at the most efficient level and ensuring fair sharing of water between sectors.

Flow Estimation and Basin Optimization Model (ATHOM) software has been developed to estimate the flow to water structures and accordingly prepare the water allocation of storage structures. In this context, two types of flow prediction models have been developed. While the first one estimates the inflows of dams in the basin for a short term with an hourly resolution of up to 10 days, the second one estimates the dam inflows for a long term with a daily resolution of up to 15 months. The cascade basin is operated for flood purposes using short-term flow forecasts, and annual operation curves for energy purposes are created using long-term flow forecasts.

ATHOM, a dynamic and operational system, takes numerical weather forecast data and automatic meteorological observation station data as input into the system. In addition, water data and energy data of dams are continuously input into the system. Flow forecasts and operating curves created with the developed stream forecasting and basin optimization models are presented to system users through a web-based application containing a geographical information system. Meteorological forecasts produced by both Europe and the model have been expanded to cover all of Türkiye.

Developed with artificial intelligence-based models, ATHOM ensures the most efficient use of water and maximum energy production with an average prediction accuracy of up to 95% in flow predictions. With correct operating policies, it is possible to produce at least 5% more energy annually (Buhan et al., 2020).

6.4.4 Implementation of Discount Fee on Use of Gradual Water and Modern Irrigation Methods

The importance of implementing the internationally accepted principles of "user pays" and "polluter pays" is becoming increasingly important with global climate change. In order to prevent the use of water more than necessary, "gradual water usage service fee" is applied in piped irrigation systems with suitable infrastructures.

In pricing based on volume, which is applied as a standard, the cubic meter price is fixed, whereas in the gradual water usage service fee application, the price increases or decreases with each increase or decrease in the amount of water used. Thus, the water usage service fee will be applied with an approach that increasingly rewards when less water is used or increasingly penalizes when too much water is used, encouraging the use of water as much as the crop needs.

On the other hand discounts are made for farmers used sprinkler and drip irrigation applications in order to support water-saving irrigation methods in the operation and maintenance fee tariffs published for irrigation facilities operated by DSI and in the water usage service fee tariffs to be taken as the minimum value in irrigation systems operated by irrigation unions.

In order to support water and energy saving, a 50% discount is applied to the irrigation fee for the crops that irrigated by sprinkler and drip irrigation systems.

6.4.5 The Drought Monitoring System

Within the scope of the Türkiye Flood and Drought Management Project, various indices were selected in order to monitor meteorological, agricultural and hydrological droughts, to determine the severity, spread and duration of possible drought events, and to reveal the spread and continuity of sudden agricultural droughts.

In this context, (i) in the monitoring activities to be carried out in the Ceyhan Basin, which was selected as a pilot to comprehensively evaluate water requirements in order to increase resilience to drought by optimizing the use of irrigation water, meteorological stations, soil water content monitoring stations, flow observation stations, lake observation stations, and other real-time data sources will be installed and (ii) the drought monitoring map information platform will be developed. This initiative aims to help farmers improve water use efficiency, diversify crop choices, and improve irrigation efficiency. This activity will build on previous activities implemented by DSI and other relevant institutions for drought monitoring.

In addition, in order to increase the success of irrigation management activities in drought conditions, irrigation timing will be planned using adjusted plant water needs specific to irrigation and parcel water distribution programs will be prepared and implemented with decision support systems. On the other hand, with artificial intelligence, applications such as preparing limited irrigation simulations that provide the least productivity loss and the highest economic return in drought conditions, giving incentives or making compulsory for plants that do not consume too much water in areas with water constraints, or making difference payments to avoid loss of income will be directed.

6.4.6 Subsidizing Pressurized Irrigation Systems

Individual irrigation systems support is provided to ensure that producers use modern pressurized individual irrigation systems developed for agricultural activities, to ensure higher quality production in line with market demands, and to increase the income level of producers in rural areas. In the support work that has been ongoing since 2007, the Ministry provides grant support of 50% of the investment made by farmers in a certain amount that is updated every year.

Within the scope of the program; Investments in the establishment of in-field drip, sprinkler, micro sprinkler, subsurface drip, linear or center pivot, drum irrigation systems, solar energy irrigation system, solar energy systems for agricultural irrigation, and smart irrigation systems are supported. Since the start of the support in 2007, more than 2 billion TL support has been provided to more than 60 thousand projects.

6.4.7   The Role of Water User Associations (WUAs)

The United Nations Food and Agriculture Organization (FAO) described irrigation modernization as: a process of technical and managerial upgrading (as opposed to mere rehabilitation) of irrigation schemes combined with institutional reforms, with the objective of improving resource utilization (water, labor, economic, and environmental) and water delivery to farms. (Renault et al., 2007).

In this respect, WUAs are, thus, one of most important key factor for the efficient use of water in agriculture. Transfer of operation, maintenance, and management responsibilities of irrigation systems from DSI to WUAs has gained momentum since 1993 in parallel with the conjuncture in the world. While small and isolated projects were being transferred before—the first transfer goes back to the 1930s—transfer activities began to include large-scale irrigation systems after 1993. In current situation, in terms of operation and maintenance responsibility of irrigation systems, almost 85% of it transferred to WUAs. For this reason, the capacity of WUAs are very related to water efficiency.

Türkiye made a big amendment in the management of WUA's in 2018 by law number 7139. Elected president, board members and council of WUA's canceled and a government officials have been attended as president, taking all responsibilities belonging to old management team of WUA's. It may be seen as strange for some. However, countries can develop new administrative amendments in accordance with their special conditions. On the other hand, water is the most strategic resource for living, so, very close control of the government in irrigation management is necessary.

After amendment, new and strict audits rules have begun to prevent misuse. WUA's have begun managing by competent engineers. The number of 384 WUA's decreased 181 by having united them. It results in more economic administrative and financial results, because of scale economy.

Currently, WUA's have made good technical, financial, and administrative progresses. These progresses can be summarized as; reduction in O&M expenditures, financial discipline, efficient use of water, safe water transmission, more equitable, reliable, and adequate fair water distribution, reliable and accountable management, fast and effective decision-making process, improving collaborative relations between farmers and local administrations.

6.5 Conclusion

Limited water resources must be used effectively in all sectors in harmony with the environment. Activities on the effective use of water in agriculture are of great importance in terms of the effective use of natural resources that will be protected for future generations.

Ensuring the effective use of limited water resources in the agricultural sector starts from the water supply stage, water distribution and control, maintenance-repair and renewal of irrigation systems, monitoring and evaluation of irrigation systems, planning and development of irrigation projects, deciding on irrigation costs, resolving disputes between farmers, determining plant patterns. It creates a very wide chain of activities such as planning and creating an irrigation calendar. Water resources supply, use and management are evaluated with a holistic approach in Türkiye.

In order to irrigate a larger area with existing water resources, it is of great importance to select the most appropriate irrigation method by considering basic conditions, development of existing irrigation technologies, soil, plant, water source, economy and similar factors, and to establish and operate the irrigation system required by the method.

In order to save irrigation water and benefit more from unit water, it is not enough to take only structural measures such as popularizing pressurized pipe irrigation systems, establishing the necessary infrastructure to ensure moderate use of water at all levels, popularizing automation and decision support systems, and carrying out maintenance and repair. Water users should also be encouraged to make changes in their water usage habits. In this context, special efforts are being made to ensure that farmers prefer sprinkler or drip irrigation methods in their field applications.

No doubt that modernization of irrigation for efficient water use is not only consist of the structural Works mentioned above. It includes institutional and administrative dimensions. In this respect WUA's are the best model. However, every country should adapt administrative formation of WUA's to its own more suitable one.

References

Adali, S., & Kiliç, M. Y. (2020). Use of treated wastewater in agricultural irrigation: Iznik example. Uluslararasi Biyosistem Mühendisliği Dergisi (International Journal of Biosystems Engineering), *1*(1), 12–23. (in Turkish).

Anonymous. (2017). 2nd Forestry and water council. Irrigation Working Group Document. 31. (in Turkish).

Armstrong, M. (2024). Where water stress will be highest by 2050. Retrieved from https://www.statista.com/chart/26140/water-stress-projections-global/ (Access date: 7 July 2024).

Buhan, S., Küçük, D., Çınar, M. S., Güvengir, U., Demirci, T., Yilmaz, Y., Malkoç, F., Eminoğlu, E., & Yildirim, M. U. (2020). A scalable river flow forecast and basin optimization system for hydropower plants. *IEEE Transactions on Sustainable Energy, 11*(4), 2220–2229.

Çetin, Ö., Fayrap, A., & Yolcu, R. (2024). Sustainability and modernization of agricultural irrigation: A comparative assessment of two irrigation schemes. *Irrigation and Drainage, 73*(1), 284–293. https://doi.org/10.1002/ird.2878

DSI (2020). Basin Master Plans. Executive Summary. Department of Survey, Planning and Allocations of DSI, Ankara, Türkiye. (in Turkish).

DSI. (2024a). Soil water resources. Retrieved from https://dsi.gov.tr/Sayfa/Detay/754 (Access date: 7 July 2024) (in Turkish).

DSI. (2024b). 2023 (in Turkish) (Assessment report of 2023 in irrigation schemes operated by DSI). İşletme ve Bakim Dairesi Başkanliği Yayinlari (Operation and Maintenance Department), Ankara, Türkiye.

FAO. (2023). World Food and Agriculture: Statistical Yearbook 2023, Rome.

FAO. (2024). AQUASTAT-FAO's global information system on water and agriculture. Retrieved from https://www.fao.org/aquastat/en/overview/methodology/water-use (Access date : 7 July 2024).

Kuzma, S., Bierkens, M. F. P., Lakshman, S., Luo, T., Saccoccia, L., Sutanudjaja, E. H., & Van Beek, R. (2023). Aqueduct 4.0: updated, decision-relevant global water risk indicators, technical note. World Resources Institute, Washington, DC. https://doi.org/10.46830/writn.23.00061

Natalizio, M. (2015). A modern management model for irrigation systems. In: *Proceedings of sixth international scientific agricultural symposium "Agrosym 2015"* (pp. 1338–1346). Jahorina, 15–18 October, Bosnia and Herzegovina, ISBN 978-99976-632-2-1, COBISS.RS-ID 5461016, https://doi.org/10.7251/AGSY15051338N

Perry, C., Steduto, P., & Karajeh, F. (2017). Does improved irrigation technology save water? A review of the evidence discussion paper on irrigation and sustainable water resources management in the Near East and North Africa. FAO Technical Report May 2017. https://doi.org/10.13140/RG.2.2.35540.81280

Renault, D., Facon, T., & Wahaj, R. (2007). Modernizing irrigation management-MASSCOTE approach I&D Paper 63. FAO, Rome.

Şener, M., Erdem, T., Çelen, H., Tekiner, M., Yildirim, M. U., Pehlivan, M., Şeren, A., Kolsuz H. U., Seyrek, K., & Turan, L. (2020). Determination of irrigated parcels using unmanned aerial vehicles and remote sensing technology. *DSI Technical Bulletin* (136), 20–36.

Şeren, A., & Çarbaş, A. (2015). Sustainability of irrigation facilities. *Water World* (147), 46–53. Retrieved from https://cdniys.tarimorman.gov.tr/api/File/GetGaleriFile/425/DosyaGaleri/1010/147.pdf (Access date : 3 July 2024).

Şeren, A. (2023). Decision support systems in irrigation management: Irrigation facilities spatial information system. *Irrigation and Drainage*, 1–9. https://doi.org/10.1002/ird.2808

TRGM (2022). Agricultural drought combat strategy and action plan for Türkiye, 2023–2027. 17, Ankara, Türkiye. (in Turkish).

UN. (2023). The sustainable development goals report-special edition. Towards a Rescue Plan for People and Planet. 24–25. Retrieved fom https://unstats.un.org/sdgs/report/2023/Goal-06/ (Access date: 12 June 12, 2024).

WWAP. (2017). The United Nations World Water Development Report 2017. Wastewater: The Untapped Resource. Paris, UNESCO. Retrieved from http://www.unesco.org/open-access/terms-use-ccbysa-en (Access date: 10 June 2024).

Yildirim, M. U., & Gül, A. (2018). A research on social and economic aspects of waste water irrigation and alternative approaches: Case study in Gaziantep), Ç.Ü. *Journal of Science and Engineering Sciences, 35*(9), 97–106.

Chapter 7
Energy, Economic Growth, and Climate Change in Southeast European Countries

Flora Merko⬭ and Ermira H. Kalaj⬭

Abstract This study uses annual time series data from 2000 to 2021 to investigate the relationship between primary energy consumption and growth in Albania, Serbia, Croatia, Bulgaria, and Romania. These countries are economies that are located in Southern Eastern Europe and have been through a number of episodes that have made them of particular importance to the study of periods of economic expansion and stagnation. There is an extensive body of research that focuses on the relationship between the consumption of energy and the expansion of the economy. It is fact that climate change will also affects all these. The relationship between energy and growth has significant implications from various perspectives, including theoretical, practical, and policy considerations. The real gross domestic product per capita is the dependent variable that we employ in our estimated model. The independent variables that we have selected are as follows: energy consumption per capita, real gross fixed capital formation, exports, imports, and the political instability index among others. According to the findings of our estimation, political instability is a factor that hinders the process of economic growth and development in some countries. This finding is in line with our earlier expectations.

Keywords Climate change · Energy consumption · Economic growth · European countries

7.1 Introduction

The relationship between energy and economic growth is a crucial topic in economics and sustainability studies. There are some key points to understand this relationship, which are related to some characteristics of energy and relations it has with

F. Merko (✉)
University "Aleksandër Moisiu", Durrës, Albania
e-mail: floramerko@uamd.edu.al

E. H. Kalaj
University "Luigj Gurakuqi", Shkodër, Albania
e-mail: ermira.kalaj@unishk.edu.al

economic growth. Energy serves as a driver of economic growth, which relates with industrialization, increasing productivity and efficiency of agriculture and water use, transportation and mobility, where energy has been a fundamental driver, allows for greater productivity and efficiency, and is essential for trade and the movement of goods and people. Energy is also related to environment, because climate changes are driving a transition to cleaner, renewable energy sources. Energy is considered an essential input for economic activities, and increased energy consumption often accompanies industrialization and economic development (Stern, 2000). According to Sari et al. (2008), Sharma (2010) and Magazzino (2012), energy is crucial within supply chains. It serves as a non-durable consumer good and is essential as an input in the production processes of various types of businesses. Both consumers and businesses worldwide consider energy to be one of the most crucial commodities ever provided. An enormous body of research has been conducted to establish a connection between economic growth and energy consumption. However, the empirical findings have generated results that are equivocal on the direction of the causality effect.

The agriculture sector might benefit greatly from renewable resources of energy. Subsidies should be provided to farmers to stimulate the usage of renewable energy technologies. Sustainable agriculture is based on a careful balancing act between reducing the use of limited natural resources and negative environmental effects and optimizing crop output and preserving economic stability. Restoring the soil while using as little non-renewable resource as possible is also essential to sustainable agriculture. Examples of these resources include natural gas, which is used to turn atmospheric nitrogen into synthetic fertilizer, mineral ores like phosphate, and fossil fuels like diesel generators, which are used to pump water for irrigation. Therefore, it is imperative to encourage the use of renewable energy systems for sustainable agriculture, such as solar hot water heaters, greenhouse technologies, solar dryers for post-harvest processing, and solar photovoltaic water pumps and electricity.

The consumption of energy was, thus, found to be an effective factor in promoting economic growth in a number of empirical researches for all sectors such as industry, agriculture, and daily life (Alper, 2016; Apergis & Tang, 2013; Chen et al., 2020; Gyimah et al., 2022). Other empirical research, on the other hand, has put forth the argument that economic expansion is not the outcome of increased energy consumption (Huanget et al. 2008; Shahbaz & Feridun, 2012; Acheampong, 2018).

Economic growth is a fundamental component of macroeconomic study. Hence, it is crucial to identify the true drivers of growth in order to develop efficient policy instruments that will facilitate sustainable growth in the economy. In theory, energy plays an essential role in promoting economic growth. In addition, Stern (2011) observed that energy influences economic growth by affecting production activities.

The relationship between primary energy consumption and economic growth is a well-researched topic also for the countries we choose for study such as, Albania, Serbia, Croatia, Bulgaria, and Romania. However, the empirical evidence that looks into the link between energy and growth, the findings, are not the same. Chaudhry et al. (2012), point that energy use is a good determinant of economic growth while other studies find insignificant (Ozturk et al., 2010), bi-directional, or even negative effects (Narayan, 2016; Yildirim et al., 2014). Thus, many studies across different countries have found a positive relationship between energy consumption and economic growth, especially in developing and transition economies (Payne, 2010; Soytas & Sari, 2003).

Muco et al. (2021) stated that the importance of energy attracts the study of the relationship between energy consumption and economic growth, since higher economic growth leads to a higher level of energy consumption. The findings of the study (for European Transition Economies), show that a growing Gross Domestic Product (GDP) implies a growing energy use, but on other side a higher use of energy is not good for the environment. Another positive effect of economic growth on energy consumption, was found in the study carried out by Mombekova et al. (2024) where some developing countries (China, India, South Africa, Indonesia, Turkiye, Mexico, Thailand) are studied.

The causality result for Albania was showed by Kumar et al., 2019, that economic growth drives energy consumption, and the elasticity of income with respect to energy is 0.36, showing that, ceteris paribus, a 1% increase in energy consumption will increase output by 0.36%. Çetintaş (2016) has investigated the causality relationship between energy consumption and economic growth in 17 transition countries including Albania. Empirical findings indicate that there is unidirectional causality from economic growth to energy consumption in the long run, suggesting that these countries can simultaneously achieve policy goals concerning growth and energy. Obradović and Lojanica (2017) have studied causal relations among economic growth, energy, and emissions on the example of Greece and Bulgaria. In the long-run results suggest that energy is one of the engines of growth, but for the short-run results the opposite for both countries.

The post communists' countries (Albania, Bulgaria, Romania, and also Croatia and Serbia as part of ex-Yugoslavia) have been in focus of Umurzakov et al. (2020), linked with energy consumption and economic growth in these countries. Panel Dynamic OLS results revealed a positive and significant impact of energy consumption on economic growth in post-communistic countries, and on the other side, economic growth causes energy consumption confirming the conservation hypothesis.

In this chapter, it is reviewed that the relationship between primary energy consumption and economic growth focused on Southeast European Countries such as Albania, Serbia, Croatia, Bulgaria, and Romania, and more, trying to show that the relationship between energy and economic growth is complex and multifaceted. While energy is a critical driver of economic development, ensuring sustainable growth requires a balanced approach that considers energy efficiency, environmental impact, and technological innovation.

7.2 Background of the Energy Market in Some European Countries

The energy markets in Albania, Serbia, Croatia, Bulgaria, and Romania have unique characteristics shaped by their histories, resources, and policy environments (Figs. 7.1 and 7.2).

Albania is characterized by hydropower dominance. Albania's energy market is heavily reliant on hydropower, which accounts for almost all of its electricity generation. Major hydropower plants on the Drin River supply the bulk of the country's energy needs. During periods of drought, Albania faces energy shortages and relies on imports to meet demand. Albania has undertaken significant reforms to liberalize its energy market, improve regulatory frameworks, and attract foreign investment. Albania is, on one hand, an energy-resource abundant country, and on the other hand, it was not able to cover the domestic electricity demand in the recent years. Albania is the largest producer of crude oil in Europe. State-run oil firm AlbPetrol estimates that Albania has recoverable oil reserves of around 120 million barrels, and natural gas reserves of 5.7 billion m^3 gas (ITA, 2021). The Patos-Marinza oilfield is the largest onshore oil field in continental Europe. In 2023, available electricity decreased by 0.6% compared to the previous year. The net domestic production of electric power during this period amounted to 8796 GWh, a notable increase from 7003 GWh in 2022, marking a rise of 25.6%. Public hydro plants accounted for 58.2% of the net domestic production, while independent power producers contributed 40.8%, and other producers, such as photovoltaics, generated 1.0%. Gross imports of electric power (including exchanges) totaled 1922 GWh, down by 36.9% from the previous year's 3044 GWh. Gross exports (including exchanges) amounted to 2842 GWh, indicating an increase of 33.9% from 2123 GWh in the previous year (INSTAT, 2024).

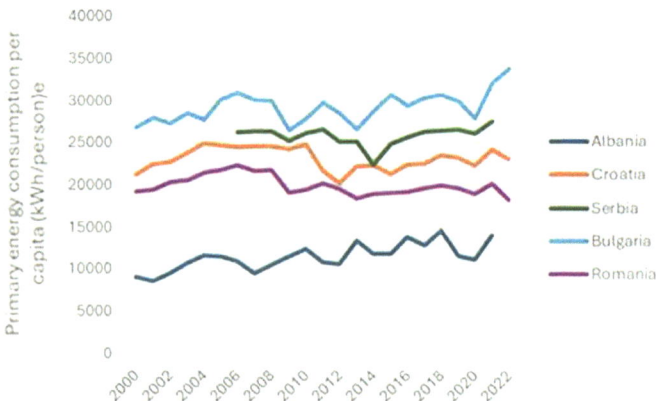

Fig. 7.1 Primary energy consumption per capita for the countries

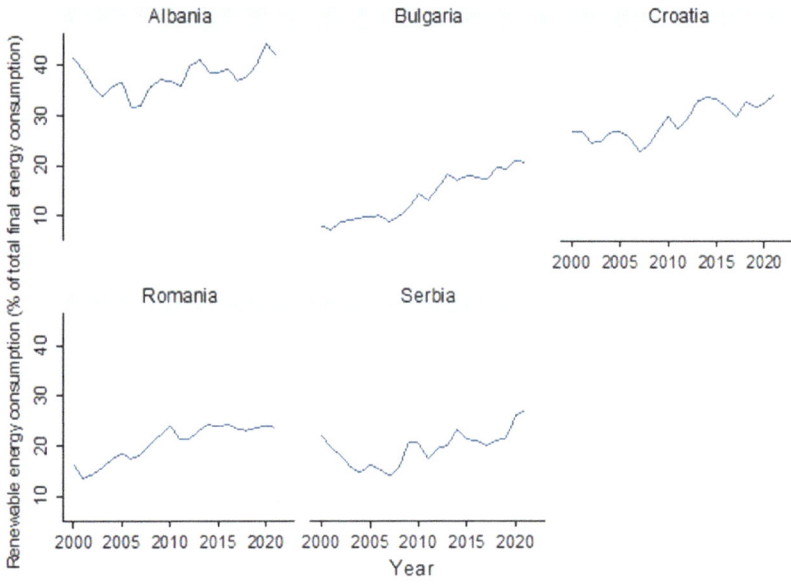

Fig. 7.2 Renewable energy consumption as % of final energy consumption, by country

One of the main obstacles of Albania's energy sector is the missing reliability and low profitability of the electricity sector. Bidaj et al (2015) estimate that 29–52% of the total electricity supply in Albania, is lost due to technical and non-technical (theft and unpaid bills) reasons. In the condition of this sector, the energy infrastructure requires upgrades to improve reliability and reduce losses in the transmission and distribution network. There is a need to diversify energy sources to enhance energy security and reduce vulnerability to hydrological variability. Albania is increasingly focusing on renewable energy sources to meet its energy needs.

Serbia's energy mix is dominated by coal, particularly lignite, which fuels most of its electricity generation.

Hydropower is the second-largest source of electricity. The country is also exploring the potential of renewable energy sources like wind and solar. Serbia is gradually liberalizing its energy market, aiming to align with EU regulations and attract investment. The heavy reliance on coal poses significant environmental and health issues due to high greenhouse gas emissions. Modernizing the aging energy infrastructure and increasing investment in renewable energy are critical. The National Action Plan for Renewable Energy (NREAP, 2013) aimed for 27% of renewables in final energy consumption by 2020 (with targets of 37% for electricity, 30% for heating, and 10% for transport). Serbia narrowly missed this goal, achieving a 26% share in 2020. By 2022, the proportion dropped further to below 25% (with 30% in electricity, 35% in heating, and 0.6% in transport). The 27% target has now been rescheduled for achievement by 2025 (Anonymous, 2024a).

Croatia has a relatively diverse energy mix, including natural gas, hydropower, and renewables. It also imports electricity to meet demand. As an EU member, Croatia's energy market is integrated with the EU's energy policies and regulations. Croatia is actively promoting renewable energy projects, particularly in wind and solar power. Continued efforts are needed to fully liberalize the energy market and foster competition. Upgrading energy infrastructure and improving grid connectivity are ongoing priorities. Compared with Serbia, in 2022, Croatia achieved a renewable energy coverage of 28% in final energy consumption, distributed as follows: 55.5% for electricity, 37% for heating and cooling, and 2.4% in transport. This performance significantly surpassed the 2020 target set by the National Renewable Energy Action Plan (NREAP, 2013), exceeding it by 11 percentage points (Anonymous, 2024b). Bulgaria has a significant nuclear power capacity, with the Kozloduy Nuclear Power Plant providing a substantial portion of its electricity. Coal remains an important energy source, but the country is also increasing its investment in renewable energy, especially wind and solar. Bulgaria exports electricity to neighboring countries, benefiting from its strategic location in Southeast Europe. The total consumption of 30.91 billion kWh of electric energy per year, and 4780 kWh per capita, are the most important figures in the energy balance of Bulgaria, which could be self-sufficient with domestically produced energy. The reliance on coal poses environmental challenges, and there is pressure to reduce emissions in line with EU targets. Enhancing the regulatory framework to attract investment in renewables and modernize the energy sector is essential. In 2021, renewable energy accounted for around 20.4 percent of actual total consumption in Bulgaria (Anonymous, 2024c).

Romania has a diverse energy mix, including natural gas, coal, nuclear, hydropower, and renewables. The country has substantial oil and natural gas reserves, making it relatively energy independent compared to its neighbors. As an EU member, Romania aligns its energy policies with EU directives, promoting market liberalization and sustainability. Upgrading aging energy infrastructure and improving energy efficiency are ongoing challenges. Increasing the share of renewable energy in the energy mix and ensuring grid stability are key priorities. The total consumption in Romania is pretty higher then in Bulgaria, 50.04 billion kWh of electric energy per year, but the consumption for capita is 2627 kWh. In Romania in 2021, renewable energy accounted for around 23.6% of actual total consumption (Anonymous, 2024d).

In conclusion, it could be stated that the energy markets in Albania, Serbia, Croatia, Bulgaria, and Romania are at different stages of development and face distinct challenges. Common themes include the need for infrastructure modernization, diversification of energy sources, environmental sustainability, and alignment with EU energy policies. Each country is working to balance energy security, economic growth, and environmental protection in its energy strategies.

7.3 Data and Methodology

In this chapter, two different sources of data were used, the World Development Indicators (WB, 2024) and International Monetary Fund (IMF, 2024), and the annual energy consumption data, collected from the database Our World in Data (OWID) for the years, 2000–2022, was obtained by Ritchie et al. (2024). The dependent and independent variables are explained in Table 7.1.

Table 7.1 Description of variables

Dependent variable	Source	Description
GDP per capita	WDI	GDP per capita is gross domestic product divided by midyear population. GDP is the sum of gross value added by all resident producers in the economy plus any product taxes and minus any subsidies not included in the value of the products
GDP growth		Annual percentage growth rate of GDP at market prices based on constant local currency. Aggregates are based on constant 2015 prices, expressed in U.S. dollars
Explanatory variables		*Description*
Renewable energy	WDI	Renewable energy consumption is the share of renewable energy in total final energy consumption
Gross fixed capital		Gross fixed capital formation (formerly gross domestic fixed investment) includes land improvements (fences, ditches, drains, and so on); plant, machinery, and equipment purchases; and the construction of roads, railways, private residential dwellings, and commercial and industrial buildings
Import of goods		Imports of goods and services represent the value of all goods and other market services received from the rest of the world
Export of goods		Exports of goods and services represent the value of all goods and other market services provided to the rest of the world
Primary energy consumption	Ourworldindata	Primary energy consumption (TWh)
Primary energy consumption per capita		Primary energy consumption per capita (kWh/person)
Political stability	IMF	Political stability and absence of violence/terrorism: estimate
Interest rate	IMF	Real effective exchange rate is the nominal effective exchange rate (a measure of the value of a currency against a weighted average of several foreign currencies) divided by a price deflator or index of costs

In this study, it is used a Linear dynamic panel-data model to identify the nexus between economic growth and energy consumption. For the parameters of this model, Arellano and Bond (1991) developed a consistent generalized method of moments (GMM) estimator, which is implemented by xtabond (Arellano–Bond estimator). This estimator demands that the idiosyncratic errors have no autocorrelation and are intended for datasets with many panels and short durations. For a similar estimator that does not require autocorrelation in the idiosyncratic errors but makes use of additional moment requirements.

To investigate the impact of energy consumption and give answer to our research questions we use the following model:

$$Y_i = \beta_0 + \beta_1 \text{Energy consumption}_i + \gamma X_i + \mu_i \qquad (7.1)$$

where: Y_i is one of the components of economic growth measured in terms of GDP per capita, and GDP annual growth.

Energy consumption, are variables to indicate the consumption of energy either in terms of *primary energy per capita* or *renewable energy* by country.

X_i is vector of variables including gross fixed capital formation, import and export of goods, political stability, and real interest rate.

After determining that the variables primary energy consumption per capita and renewable energy are significantly associated, as indicated by the matrix of correlations presented in Table 7.2, we continue with performing an array of regressions for these independent variables.

According to the coefficients of the correlation, the estimated coefficient is more than 0.80 only in the case of primary energy consumption per capita and the variable renewable energy. This is something that we can observe from the coefficients. For this purpose, we will conduct different analyses to assess the impact that renewable energy has had on the economic growth of the countries that have been chosen.

7.4 Empirical Results

The Southeast European countries that have been selected for the sample are different from one another in terms of their level of economic development and their energy consumption; yet they are all part of the group of Balkan countries. While Albania and Serbia are not yet members of the European Union, some of them are already members of the EU. The econometric results of the regression for GDP per capita growth are shown in Table 7.3. Taking into the coefficients account, the only statistically significant are the lagged GDP per capita, gross fixed capital, and primary energy consumption per capita. The later result is in line with previous studies conducted in other countries (Acaravci & Ozturk, 2010; Aydin, 2019; Bozoklu & Yilanci, 2013; Triantafyllidou et al., 2023). However, in terms of magnitude the effect is not so high. In the sample for the period 2000–2021 we investigate the link between gross domestic product and primary energy consumption. One may observe

Table 7.2 Matrix of correlation

Variables	(1)	(2)	(3)	(4)	(5)	(6)	(7)	(8)	(9)	(10)
(1) GDP per capita	1.000									
(2) GDP growth	−0.249	1.000								
(3) Renewable energy	0.059	−0.120	1.000							
(4) Gross fixed capital	−0.529	0.530	0.370	1.000						
(5) Imports of goods	0.245	0.071	−0.121	−0.004	1.000					
(6) Exports of goods	0.406	−0.092	−0.347	−0.400	0.783	1.000				
(7) Primary energy (kwh)	0.073	0.097	−0.656	−0.317	−0.287	0.128	1.000			
(8) Primary energy. cons/cap	0.337	−0.014	−0.801	−0.573	0.397	0.626	0.362	1.000		
(9) Political stability	0.651	−0.213	−0.038	−0.421	0.147	0.338	0.020	0.354	1.000	
(10) Interest rate	0.628	−0.166	0.236	−0.274	−0.015	−0.055	−0.340	0.139	0.627	1.000

that a percentage rise in the use of primary energy results in a comparable increase of 0.25% in the gross domestic product.

According to the coefficients in Table 7.4, the estimated coefficient for the primary energy consumption per capita is still statistically significant but the magnitude or effect is quite lower when the dependent variable used is GDP growth. Moreover, another significant relationship comes to the light, the one between the economic growth and imports of goods and services.

The effect of renewable energy on economic growth is shown in Table 7.5. In this table the estimations are shown by country. As expected in all the economies selected by our study the lag of GDP growth positively impacts economic growth. Gross fixed capital formation still robustly affects economic development. The estimated coefficients for renewable energy are positive in the cases of Albania, Croatia, and Bulgaria, while the relationship is negative in the estimations for Serbia and Romania. The negative relationship is in accordance with the conservation hypothesis, and the neutrality hypothesis (Aydin, 2019). On the other hand, to acquire a deeper

Table 7.3 Arellano-Bond dynamic panel-data estimation on GDP per capita, for the whole sample

GDP per capita	Coef	St. Err	t-value	p-value	[95% Conf	Interval]	Sig
L. GDP per capita	0.949	0.051	18.47	0	0.848	1.05	***
Gross fixed capital	27.628	10.733	2.57	0.01	6.593	48.664	**
Imports of goods	38.907	24.329	1.60	0.11	−8.777	86.592	
Exports of goods	−11.086	19.336	−0.57	0.566	−48.984	26.813	
Primary energy cons	0.25	0.066	3.82	0	0.122	0.379	***
Political stability	218.918	487.138	0.45	0.653	−735.855	1173.691	
Interest rate	−48.773	40.973	−1.19	0.234	−129.078	31.533	
Constant	−4971.977	1740.476	−2.86	0.004	−8383.248	−1560.706	***
Mean dependent var	8330.607		SD dependent var	4318.209			
Number of obs	83.000		Chi-square	1046.351			

*** $p < 0.01$, ** $p < 0.05$, * $p < 0.1$

Table 7.4 Arellano-Bond dynamic panel-data estimation on GDP growth, for the whole sample

GDP growth	Coef	St. Err	t-value	p-value	[95% Conf	Interval]	Sig
L.GDP growth	−0.121	0.091	−1.33	0.183	−0.3	0.057	
Gross fixed capital	0.168	0.034	4.96	0	0.101	0.234	***
Imports of goods	0.159	0.09	1.77	0.077	−0.017	0.335	*
Exports of goods	−0.087	0.071	−1.22	0.224	−0.226	0.053	
Primary energy cons	0.001	0	3.33	0.001	0	0.001	***
Political stability	−2.211	1.724	−1.28	0.2	−5.59	1.167	
Interest rate	−0.144	0.143	−1.01	0.312	−0.424	0.136	
Constant	−14.926	6.197	−2.41	0.016	−27.071	−2.78	**
Mean dependent var	3.265		SD dependent var		3.789		
Number of obs	83.000		Chi-square		80.503		

*** $p < 0.01$, ** $p < 0.05$, * $p < 0.1$

comprehension of this relationship, additional research must be performed about the definition of renewable energy and the policies that contribute to it.

Table 7.5 Arellano-Bond dynamic estimation on GDP growth, by country

GDP growth	(1)	(2)	(3)	(5)	(4)
	Albania	Croatia	Serbia	Bulgaria	Romania
L.GDP growth	0.454*	0.943***	0.775***	0.363**	0.001
	(0.235)	(0.276)	(0.22)	(0.167)	(0.168)
Gross fixed capital	0.938***	0.199***	0.427***	0.327***	−0.012
	(0.259)	(0.068)	(0.108)	(0.101)	(0.075)
Imports of goods	−0.104	−0.123	0.344	0.431*	1.333**
	(0.211)	(0.161)	(0.245)	(0.234)	(0.53)
Exports of goods	0.569**	0.35	−0.137	−0.147	−0.179
	(0.257)	(0.215)	(0.242)	(0.202)	(0.28)
Renewable energy	0.014	1.392***	−0.649	0.329**	−1.496**
	(0.273)	(0.478)	(0.494)	(0.152)	(0.713)
Political stability	4.859*	−1.53	6.161	−2.457	−6.434*
	(2.545)	(3.749)	(10.912)	(10.234)	(3.854)
Interest rate	−0.524	−2.552***	−0.271	−0.519	0.283
	(0.569)	(0.982)	(0.336)	(0.353)	(0.462)
_cons	−29.767	29.209***	31.172	4.355	−11.557
	(20.144)	(9.071)	(38.754)	(32.309)	(20.495)
Observations	19	19	19	20	19
Pseudo R^2	0.32	0.26	0.15	0.23	0.31

Standard errors are in parentheses

*** $p < 0.01$, ** $p < 0.05$, * $p < 0.1$

7.5 Conclusion

This paper is focused on the analyses of the role of primary energy consumption on economic growth of a selection of Southeast European countries for the period 2000–2021. Economic growth is measured as a proxy of GDP per capita and GDP annual growth. To give answers to the research questions, the study relies on three different sources of data World Development Indicators, IMF databank, and Our World in Data (OWID). The selected countries are Albania, Croatia, Serbia, Bulgaria, and Romania.

In this study, Arellano-Bond dynamic panel-data model was used. Both by country and for the whole sample regressions have shown a positive relationship between economic growth and primary energy consumption. The growth hypothesis states that economic growth is causally correlated with energy consumption. In this instance, using energy as a complimentary input to labor and capital during production promotes economic growth. Furthermore, ensuring economic growth depends heavily on the amount of energy consumed. Based on the idea that there is a two-way causal relationship between economic growth and energy consumption, the feedback hypothesis was developed (Aydin, 2019). In this instance, policies must be evaluated in terms of enhancing environmental quality and the security of the electrical energy supply in addition to economic growth (Merko et al., 2020).

Findings show that renewable energy affects growth in different direction the economic growth when it comes to the countries in the study. When clean energy, lowering carbon emissions, and energy security are taken into account, the consumption of renewable energy should rise. In this regard, every nation ought to make prompt use of its own renewable energy resources. Government regulations and financial incentives are the best instruments to implement such a form of strategy.

Taking into consideration these findings, it is of the most crucial significance that additional efforts be made to speed up the transition toward renewable energy. It is possible that this will involve the implementation of rules that promote the efficient utilization of renewable energy sources in an array of sectors the strengthening of research and development activities in clean energy technology, and the provision of further financial incentives and subsidies for projects related to renewable energy.

The countries that are being analyzed should take into consideration the policy implications for the diversification of energy sources to promote sustainable development and reduce dependence on non-renewable energy sources. This can be achieved by increasing the proportion of renewable energy sources in the overall energy mix through the implementation of energy efficiency measures, developing a favorable regulatory framework that encourages investment in renewable energy sources, and providing financial incentives for investments in renewable energy infrastructure.

References

Acaravci, A., & Ozturk, I. (2010). Electricity consumption-growth nexus: Evidence from panel data for transition countries. *Energy Economics, 32*(3), 604–608. https://doi.org/10.1016/j.eneco.2009.10.016

Acheampong, A. O. (2018). Economic growth, CO2 emissions and energy consumption: What causes what and where? *Energy Economics, 74*, 677–692. https://doi.org/10.1016/j.eneco.2018.07.022

Alper, A. (2016). The role of renewable energy consumption in economic growth: Evidence from asymmetric causality. *Renewable and Sustainable Energy Reviews, 60*, 953–995. https://doi.org/10.1016/j.rser.2016.01.123

Anonymous. (2024a). Renewable in % electricity production. Retrieved from https://www.enerdata.net/estore/energy Serbia Energy Information (Access date: 24 July 2024).

Anonymous. (2024b). Renewable in % Electricity production. Retrieved from https://www.enerdata.net/estore/energy Croatia Energy Information (Access date: 24 July 2024).

Anonymous. (2024c). Usage of renewable energies. Retrieved from https://www.worlddata.info/europe/ Bulgaria energy-consumption (Access date: 24 July 2024).

Anonymous. (2024d). Usage of renewable energies. Retrieved from https://www.worlddata.info/europe/romania/ energy-consumption (Access date: 24 July 2024).

Apergis, N., & Tang, C. F. (2013). Is the energy-led growth hypothesis valid? New evidence from a sample of 85 countries. *Energy Economics, 38*, 24–31. https://doi.org/10.1016/j.eneco.2013.02.007

Arellano, M., & Bond, S. (1991). Some tests of specification for panel data: Monte Carlo evidence and an application to employment equations. The Review of Economic Studies, 58(2), 277–297. http://www.jstor.org/stable/2297968?origin=JSTOR-pdf

Aydin, M. (2019). Renewable and non-renewable electricity consumption–economic growth nexus: evidence from OECD countries. Renewable energy, 136, 599–606, https://doi.org/10.1016/j.ren ene.2019.01.008

Bidaj, F., Alushaj, R., Prifti, L., & Chittum, A. (2015). Evaluation of the heating share of household electricity consumption using statistical analysis: A case study of Tirana, Albania. *International Journal of Sustainable Energy Planning and Management, 5*, 3–14. https://doi.org/10.5278/ijs epm.2015.5.2

Bozoklu, S., & Yilanci, V. (2013). Energy consumption and economic growth for selected OECD countries: Further evidence from the Granger causality test in the frequency domain. Energy Policy, 63(C), 877–881. https://doi.org/10.1016/j.enpol.2013.09.037

Çetintaş, H. (2016). Energy consumption and economic growth: The case of transition economies. *Energy Sources, Part B: Economics, Planning, and Policy, 11*(3), 267–273. https://doi.org/10. 1080/15567249.2011.633595

Chaudhry, I. S., Safdar, N., & Farooq, F. (2012). Energy consumption and economic growth: Empirical evidence from Pakistan. *Pakistan Journal of Social Sciences, 32*(2), 371–382.

Chen, C., Pinar, M., & Stengos, T. (2020). Renewable energy consumption and economic growth nexus: Evidence from a threshold model. *Energy Policy, 139*, 95–112. https://doi.org/10.1016/ j.enpol.2020.111295

Gyimah, J., Yao, X., Tachega, M. A., Hayford, I. S., & Opoku-Mensah, E. (2022). Renewable energy consumption and economic growth: New evidence from Ghana. Energy(248C), 123559. https:// doi.org/10.1016/j.energy.2022.123559

Huang, B. N., Hwang, M. J., & Yang, C. W. (2008). Causal relationship between energy consumption and GDP growth revisited: A dynamic panel data approach. *Ecological Economics, 67*(1), 41–54. https://doi.org/10.1016/j.ecolecon.2007.11.006

INSTAT. (2024). Balance-of-electric-power 2023. Retrieved from https://www.instat.gov.al/en,env ironment-and-energy/balance-of-electric-power-2023 (Access date: 16 July 2024).

IMF. (2024). IMF Data. International Monetary Fund. Retrieved from https://www.imf.org/en//Data (Access date: 13 July 2024).

ITA. (2021). International trade administration, Albania Oil and Gas. Retrieved from https://www. trade.gov/country-commercial-guides/albania-oil-and-gas (Access date: 24 July 2024).

Kumar, R., & R., Stauvermann, P., J., & Kumar, N. (2019). Exploring the Effect of Energy Consumption on the Economic Growth of Albania. *Engineering Economics, 30*(5), 530–543. https://doi. org/10.5755/j01.ee.30.5.20177

Magazzino, C. (2012). On the relationship between disaggregated energy production and GDP in Italy. *Energy and environment, 23*(8), 1191–1207.https://doi.org/10.1260/0958-305X.23.8.1191

Merko, F., Kalaj, E., & Merko, F. (2020). How does Economic Growth Affect Deforestation-Evidence from Albania. *Journal of International Environmental Application and Science, 15*(3), 152–157.

Mombekova, G., Nurgabylov, M., Baimbetova, A., Keneshbayev, B., & Izatullayeva, B., (2024). The relationship between energy consumption, population and economic growth in developing countries. International Journal of Energy Economics and Policy, 14(3), 368–374. https://doi. org/10.32479/ijeep.15614

Muco, K., Valentini, E., & Lucarelli, S. (2021). The relationships between GDP growth, energy consumption, renewable energy production and CO_2 emissions in European transition economies. International Journal of Energy Economics and Policy. 11(4), 362–373. https://doi. org/10.32479/ijeep.11275.

Narayan, S. (2016). Predictability within the energy consumption economic growth nexus: Some evidence from income and regional groups. *Economic Modelling, 54*, 515–521. https://doi.org/ 10.1016/j.econmod.2015.12.037

Obradović, S., & Lojanica, N. (2017). Energy use, CO_2 emissions and economic growth – causality on a sample of SEE countries. *Economic Research - Ekonomska Istraživanja, 30*(1), 511–526. https://doi.org/10.1080/1331677X.2017.1305785

Ozturk, I., Aslan, A., & Kalyoncu, H. (2010). Energy consumption and economic growth relationship: Evidence from panel data for low and middle income countries. *Energy Policy, 38*(8), 4422–4428. https://doi.org/10.1016/j.enpol.2010.03.071

Payne, J. (2010). Survey of the International Evidence on the Causal Relationship Between Energy Consumption and Growth. *Journal of Economic Studies., 37*, 53–95. https://doi.org/10.1108/01443581011012261

Ritchie, H., Roser, M., & Pablo, R. (2024). Energy. Retrieved from https://ourworldindata.org/energy (Access date: 14 March 2024).

Sari, R., Ewing, B. T., & Soytas, U. (2008). The relationship between disaggregate energy consumption and industrial production in the United States: An ARDL approach. *Energy Economics, 30*(5), 2302–2313. https://doi.org/10.1016/j.eneco.2007.10.002

Shahbaz, M., & Feridun, M. (2012). Electricity consumption and economic growth empirical evidence from Pakistan. *Quality & Quantity, 46*, 1583–1599. https://doi.org/10.1007/s11135-011-9468-3

Sharma, S. S. (2010). The relationship between energy and economic growth: Empirical evidence from 66 countries. *Applied Energy, 87*(11), 3565–3574. https://doi.org/10.1016/j.apenergy.2010.06.015

Soytas, U., & Sari, R. (2003). Energy Consumption and GDP: Causality Relationship in G-7 Countries and Emerging Markets. *Energy Economics., 25*, 33–37. https://doi.org/10.1016/S0140-9883(02)00009-9

Stern, D. I. (2011). The role of energy in economic growth. *Annals of the New York Academy of Sciences, 1219*(1), 26–51. https://doi.org/10.1111/j.1749-6632.2010.05921.x

Stern, D. I. (2000). A multivariate cointegration analysis of the role of energy in the US macroeconomy. *Energy Economics, 22*(2), 267–283.

Triantafyllidou, A., Polychronidoua, P., & Mantzaris, I. (2023). Primary energy consumption and economic growth: the case of Greece. Eastern Journal of European Studies, 14(2). https://doi.org/10.47743/ejes-2023-0205.

Umurzakov, U. P., Mirzaev, B., Salahodjaev, R., Isaeva, A., Tosheva, Sh (2020). Energy consumption and economic growth: evidence from post-communist countries. International Journal of Energy Economics and Policy 10 (6), S. 59–65. https://doi.org/10.32479/ijeep.10003

WB. (2024). Databank. World development indicators. Retrieved from https://databank.worldbank.org/source/world-development-indicators (Access date: 18 June 2024).

Yildirim, E., Aslan, A., & Ozturk, I. (2014). Energy consumption and GDP in ASEAN countries: Bootstrap-corrected panel and time series causality tests. The Singapore Economic Review, 59(2), 1450010. https://doi.org/10.1142/S0217590814500106

Chapter 8
Empowering Agriculture in the Face of Climate Change with Smart Solutions

Somayyeh Razzaghi

Abstract Climate change imposes significant agricultural obstacles, impacting productivity, water resources, crop, and soil health. This chapter explores these impacts and highlights the significance of innovative approaches to enhance resilience. The chapter started with a general view of climate change and its adverse effects on agriculture. It was indicated that smart technologies are crucial in reducing these effects. Some of these techniques like precision agriculture, the Internet of Things (IoT) enabled farming, data analytics, vertical farming, robotic farming, blockchain in supply chain management, smart irrigation systems, livestock monitoring, mobile applications, and farm management software are discussed for their capability to improve resource use and enhance productivity. Moreover, the chapter covers practical applications such as climate-resilient crop selection, adaptive irrigation management, drought monitoring, and climate-smart pest and disease management to enhance climate resilience in conjunction with soil health management, agroforestry, windbreaks, carbon sequestration practices, weather index insurance, and risk management strategies in detail and emphasized the importance of knowledge sharing and capacity building. Agriculture can endure climate change by utilizing these smart solutions, promoting sustainability and food security. This holistic strategy empowers farmers to deal with the challenges of a changing climate and strengthens a resilient agricultural future.

Keywords Agriculture · Climate change · Smart solutions

8.1 Introduction

With ongoing climate change, agriculture confronts unprecedented difficulties, including extreme weather events, transitioning growing seasons, and water rarity, all of which threaten food security in our fast-evolving world (Mutengwa et al., 2023).

S. Razzaghi (✉)
Faculty of Agriculture, Department of Soil Science and Plant Nutrition, Erciyes University, Kayseri, Turkiye
e-mail: srazzaghi@erciyes.edu.tr

However, amidst these challenges lies a glimmer of hope: smart solutions. This introduction acts as a gateway to explore how technology can empower agriculture amidst climate changes, offering concrete tools to enhance resilience and sustainability.

As the impacts of the climate crisis continue to bolster, it is crucial to interpret the criticality of the challenges dealing with agriculture (Nguyen et al., 2023). From interruptions in crop yields to enhanced pest pressure, the outcomes of a changing climate are extensive and meaningful (Farooq et al., 2023). By providing context on the present condition, this information lays the core framework for a more profound inquiry into how smart solutions can outline a roadmap.

Therefore, it is better to delve into the changing potential of smart solutions in agriculture. From precision farming techniques to data-informed decision-making processes innovative technologies show the potential to mitigate the impacts of climate change (Arif et al., 2020). Through real case investigations, this chapter explored how farmers can utilize smart solutions to adopt evolving environmental conditions and promote resilience in their work. However, technological limitations, accessibility concerns, and socio-economic challenges can hinder intelligent solutions for acceptance within the agricultural sector. By knowing this information and working together to conquer it, we can confirm that the benefits of technology are available to every farmer, regardless of their geographic location and supplies (Benyam et al., 2021). This knowledge can play a crucial role in understanding how smart solutions can empower agriculture when facing the climate crisis (Smith et al., 2021). By accepting these smart solutions, farmers can overcome the difficulties of climate change, and increase crop yields for future generational divisions (Brooks & Loevinsohn, 2011).

The potential of these advanced solutions can be utilized through collective action and cooperation to improve an agricultural sector that is tougher and more sustainable (Mushunje; Salvini et al., 2016). Consequently, this chapter aims to solve a gap in previous studies by collecting insights from existing research related to empowering agriculture with smart solutions under climate change conditions. Although past research has reported the need for smart solutions, there's a lack of detailed analysis of the functionality and efficacy of these innovations. By reviewing previous literature, this chapter aims to offer a clear view of the facts, introduce fields that need more in-depth study, and clarify the potential socio-economic effects of using intelligent techniques in farming. Utilizing this, we hope to prepare the path for subsequent research and practical implementation to increase agricultural resilience and sustainability in the context of climate challenges.

8.2 Understanding the Climate Change

8.2.1 Overview of Climate Change

Climate change is a complex occurrence that refers to the persistent changing of the climate design of Earth over the long term, primarily resulting from human activities (Cowie, 2012). The primary factor in climate change is the emission of greenhouse gases into the atmosphere from different activities like burning fossil fuels for energy creation, forest destruction, industrial activities, and farming functions. These greenhouse gases, (carbon dioxide (CO_2), methane (CH_4), and nitrous oxide (N_2O)), act like a blanket, holding heat from the sun and causing an increase in the average Earth temperature called global warming (Zhang et al., 2020).

The impacts of climate change are expansive and diverse, affecting ecological systems (Malhi et al., 2020), human societies (Beniston, 2010), and the economy (Wade & Jennings, 2016). The melting of polar ice caps and glaciers, by enhancing temperatures leads to a rise in sea levels, which threatens coastal communities and habitats (Kumar et al., 2021). The mentioned variation in temperature causes some weather manifestations, including hurricanes, floods, and droughts, with severe repercussions for ecosystems, farming, human networks and also reduction in wheat production (Bell et al., 2018; Çetin et al., 2022). Climate change also presents substantial risks to human health, making respiratory illnesses worse (Khan et al., 2019), heat-related illnesses (Control & Prevention 2002), and aids the distribution of vector-borne infections (Rogers & Randolph, 2006). Climate change also affected water deficiency, interrupting ecosystems and threatening the availability of essential resources for occupations (Muluneh, 2021).

Confronting climate change needs rapid and cooperative global interventions (Sathaye et al., 2006). This target is fulfilled by the actions to decrease greenhouse gas emissions by converting to green energy sources (Mac Kinnon et al., 2018), enhancing energy efficiency, and implementing policies to limit carbon emissions (Nam & Jin, 2021). Moreover, communities need adaptation strategies like building enduring systems and promoting sustainable land-use solutions to aid in dealing with the impacts of extreme climate change (Hung et al., 2016).

8.2.2 Effects of Climate Change on Agriculture

Climate change imposes an important impact on agriculture, especially on food security (Gregory et al., 2005), rural financial systems, and worldwide ways of living (Gentle & Maraseni, 2012). The temperature alterations, combined with elevated instances of weather, interrupt growing seasons and Farming procedures (Pathak et al., 2018).

By changing precipitation levels according to increasing temperatures, some regions faced prolonged growing seasons whereas others crop growth and yield.

Certain areas may encounter extended, while others approach water deficiency or floods that injure crops and reduce soil fertility (Drebenstedt et al., 2023). Therefore, climate change can alter cultivation conditions and intensify water scarcity. Changing climate conditions create warmer temperatures that are favorable habitats for pests and diseases, which can decrease crop yields (Vijai et al., 2023). To overcome this event, management practices like pesticides must be used for a safe environment and agricultural profitability (Nam et al., 2023).

As indicated above, climate change causes decreased biological diversity which is critical for pollination and soil quality. This leads to farming systems adopting this changing situation (Fig. 8.1) (Habibullah et al., 2022; Newbold et al., 2020). In Fig 8.1, we can see that the number of species in danger of disappearing is increasing worldwide. Over the last twenty years, more plants, fishes, amphibians, mollusks, birds, mammals, insects, and reptiles are facing threats. For example, from 2006 to 2015, the number of endangered fishes, mollusks, and reptiles increased by 8.0%, 8.5%, and 12.5% each year, respectively (Habibullah et al., 2022).

Heat stress and alterations in forage availability through climate change can lead to prevent the productivity of livestock (decreased meat, milk, and egg production, particularly in areas that depend on livestock farming for their earnings (Magiri et al. 2023; Thornton et al., 2021). This phenomenon intensifies food insecurity and nutritional deficiencies, poverty, and inequality, especially in communities with restricted resources (Amoak et al., 2022).

Diminished livestock and agricultural productivity resulting in climate change cause significant economic outcomes for farmers and overstrain their livelihoods due to the high expenses for adjusting to climate change and reducing its effects (Kiem & Austin, 2013). For instance, the floods in Pakistan from 2010 (as shown in Fig. 8.2) to 2014, greatly confused many people living in rural areas (Fahad & Wang, 2018).

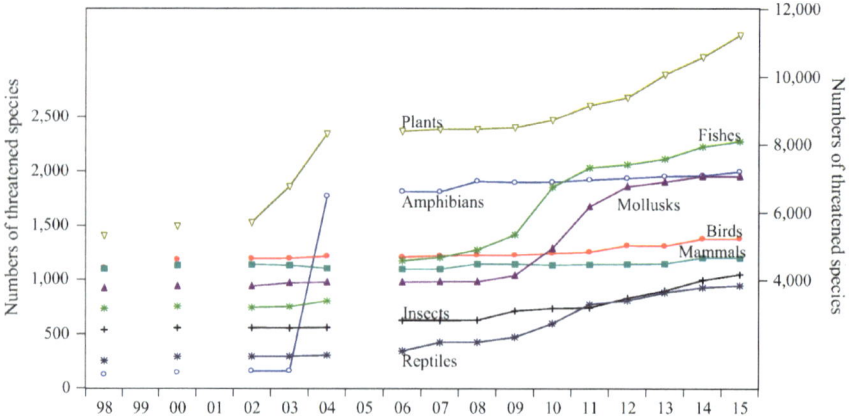

Fig. 8.1 The number of species facing threats has been going up from 1998 to 2015. (Habibullah et al., 2022)

Fig. 8.2   The flood of 2010 in the rural area of Pakistan (Fahad & Wang, 2020)

Addressing the impact of climate change on agriculture necessitates integrated action to facilitate climate-resilient agriculture practices (Singh et al., 2021), enhance farming infrastructure, optimize water management techniques (Xiong et al., 2020), and support smallholder farmers, notably in developing nations (Meemken & Bellemare, 2020). Investing in research and innovation, advocating sustainable soil management practices (Prasad et al., 2023), and promoting international collaboration is imperative for fostering climate-resilient food systems and providing food security under changing climate conditions (Ammar et al., 2023; Amoak et al., 2022).

8.3   Importance of Smart Solutions

8.3.1   Core Aspects and Advantages of Smart Technologies

Smart technologies include a diverse spectrum of innovations, including the Internet of Things (IoT), artificial intelligence (AI), machine learning, robotics, and data analytics (Khan et al., 2022; Maheswari et al. 2023). Similarly, according to Inoue (2020) as illustrated in Fig. 8.3, the 'Smart Farming' concept emphasizes four key technologies: Sensing/ GPS, information and communication technology (ICT), Big Data and AI, and Robotics. Their collaboration drives innovation in agriculture, resulting in higher productivity and quality with reduced labor and input requirements. Smart technologies can accumulate, interpret, and implement data (Sadowski, 2020) and improve effectiveness and capability enabling devices to share this information for knowledge-based decision-making, and enhanced customer observations (Sarker, 2021). Smart technologies can also offer new opportunities for innovation and sustainable advancement in different sectors (Silvestre & Ţîrcă, 2019).

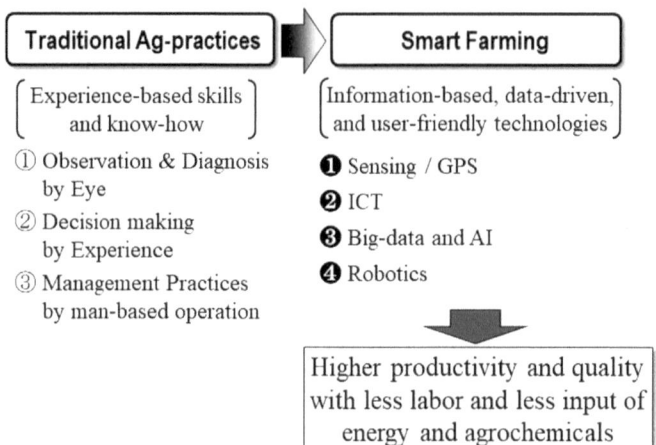

Fig. 8.3 The 'Smart Farming' concept (Inoue, 2020)

8.3.2 Benefits of Smart Technologies for Agriculture

The advantages of smart technologies for agriculture are vast and include increased efficiency (Market, 2018), improved crop yields (Andati et al., 2023), reduced resource use (Koomey et al., 2013), and enhanced decision-making (Lavorato & Piedepalumbo, 2023). These technologies enable farmers to monitor and manage their operations more effectively, leading to sustainable farming practices and greater profitability (Javaid et al., 2022). Additionally, according to Stankovic et al. (2022), smart technologies can help alleviate the effects of climate change by furnishing early warning systems for severe climatic occurrences and optimizing water usage. Overall, the acceptance of smart technologies in agriculture can revolutionize industry and address many of the challenges farmers face today.

8.3.3 Practical Applications of Smart Solutions in Agriculture

Smart solutions in agriculture utilize technology to improve productivity, sustainability, and efficiency in various aspects of farming as presented in Fig. 8.4 (Ali et al., 2023).

Precision Agriculture, IoT-enabled Farming, Data Analytics and Artificial Intelligence (AI), Vertical Farming and Controlled Environment Agriculture (CEA), Robotic Farming, Blockchain in Supply Chain Management, Smart Irrigation Systems, Livestock Monitoring and Management, Smart Irrigation Systems, Livestock Monitoring and Management, Mobile Applications and Farm Management

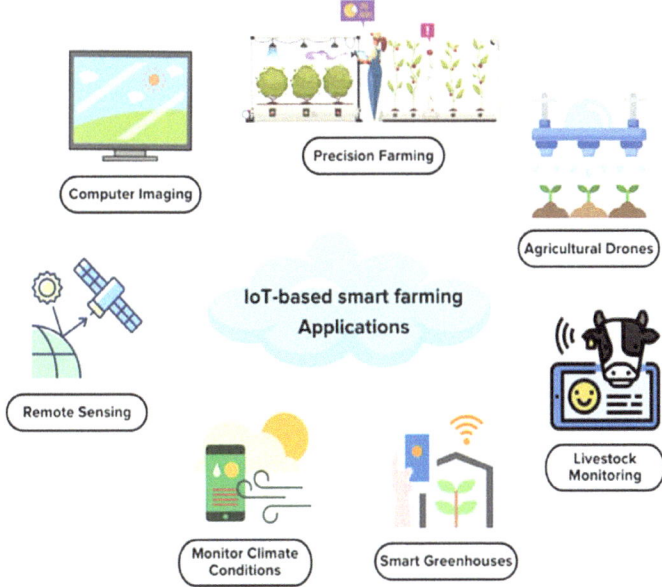

Fig. 8.4 A prime example of a central hub showcasing IoT-based smart farming applications within the realm of smart agriculture (Ali et al., 2023)

Software and Renewable Energy Integration are few practical applications of smart Technology. By using these smart applications farmers can optimize the usage of resources, and therefore, reduce environmental effects, and improve agricultural function sustainability and crop yield.

Precision agriculture utilizes GPS, drones, and sensors to observe and then manage plant growth steps like soil moisture, nutrient content, and pest invasions (Yadav & Sidana, 2023). This smart solution enables farmers to accurately manage fertilizers, pesticides, and water, resulting in low costs with high yields (Lowenberg-DeBoer, 2015).

As reported by Rajak et al. (2023) IoT appliances such as intelligent sensors and actuators can gather data on all environmental factors in real time and then analyze these results to make informed decisions to make resource use more efficient (Sinha & Dhanalakshmi, 2022).

Large datasets derived from different sources like satellite imagery, climate predictions, and historical farming data can be proceeded in advanced analytics and AI algorithms which enable predictive analytics for the estimation of crop yield, and disease detection, to help farmers make data-driven decisions (Ashraf & Akanbi, 2023; Chergui & Kechadi, 2022). CEA Indoor cultivation methods like hydroponics, aquaponics (Fig. 8.5), and vertical farming by climate control systems and smart lighting, prepare optimal growing situations and allow the plant's year-round output, optimize water consumption, and minimize land requirements (Engler & Krarti, 2021; Vatistas et al., 2022).

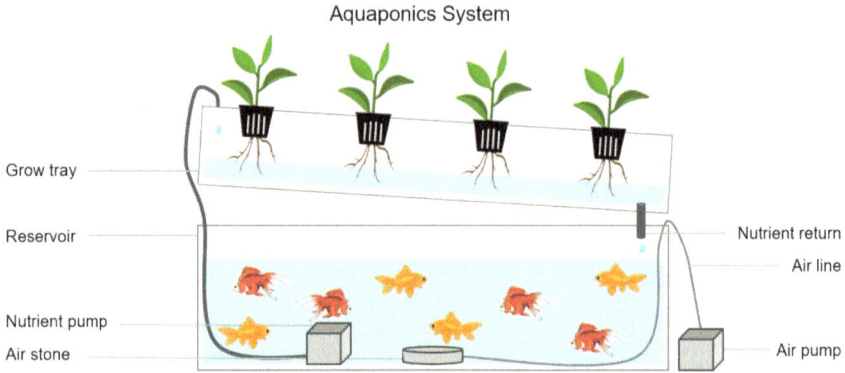

Fig. 8.5 This figure shows an aquaponics system where fish are raised to create nutrient-rich waste, which serves as food for plants in vertical farms. These plants, in turn, purify the wastewater, which is returned to the fishponds for reuse (Chole et al., 2021)

In the method of Robotic Farming according to Bechar and Vigneault (2017) and Fountas et al. (2020) Robotic systems equipped with robotic arms with cameras, and sensors, can implement planting, removing weeds, harvesting, and arranging carefully. Furthermore, the mentioned systems decrease labor costs and human error, along with increasing performance efficiency. Blockchain technology can store some data about the source and quality of a product and then can enhance the traceability of the product in the agricultural distribution network, upgrading market opportunities for farmers, and enhancing crop safety (Kshetri, 2021). To prevent overwatering and preserve water resources for optimal crop growth (Darshna et al., 2015), IoT-based irrigation systems monitor soil moisture content and climate conditions to supply water precisely when required (Salah et al., 2023).

For achieving to enhance breeding practices, prompt detection of diseases, and improve general animal welfare (Halachmi et al., 2019) wearable sensors and IoT devices are utilized to monitor the overall behavior (from health to productivity) of livestock to manage them (Alipio & Villena, 2023; Neethirajan, 2017). By Mobile Applications and Farm Management Software, farmers can provide the tools for crop planning and management, financial and market analysis, to simplify communication and office duties, and remote control of farm operations by mobile applications and software platforms (Saiz-Rubio and Rovira-Más 2020). Involvement of renewable and sustainable energy sources like wind turbines and solar panels in agricultural systems can reduce the requirement for fuel, therefore decreasing carbon emissions and indeed energy costs (Majeed et al., 2023; Pascaris et al., 2021).

8.4 The Applications of Smart Solutions in Agriculture Confronted with Climate Change

As indicated above, by smart solutions in agriculture farmers can adapt to climate change conditions by mitigating environmental effects and securing the food supply (Mutengwa et al., 2023). The applications for smart solutions in agriculture like climate-tolerant crop selection, adjustable irrigation management, water deficiency monitoring, climate-smart pest and disease management, soil quality management, agroforestry and windbreaks, carbon sequestration for climate mitigation, weather index insurance and risk management, knowledge sharing, and capacity building can help farmers confronted with climate change.

By climate-tolerant crop selection application farmers can utilize smart solutions with data-driven analyses including the use of forecasting modeling and recorded past climate data to detect crop genotypes that are more resistant to climate change for specific areas (Lal et al., 2024). In addition, by using this smart system (adjustable irrigation management) which is equipped with sensors and climate predictions farmers can irrigate plants based on real-time soil moisture levels and simultaneously reduce the effects of water deficiency without any damage to crop productivity (Chandrappa et al., 2023; Morais et al., 2005; Sasikala & Bharatula, 2024).

Furthermore, by water deficiency monitoring and early warning systems, the information gathered from satellite imagery and drones by remote sensing technologies can observe plant and soil health through moisture levels to forecast probable drought stress and alert farmers to irrigate plants on time to reduce yield losses (Inoue, 2020). Figure 8.6 presents the spectral bands of imaging techniques applicable for observing plant reactions to water scarcity, as reported in the study of Le et al. (2023).

Moreover, anticipatory modeling and sensor systems as samples of integrated pest control strategies by smart technologies (climate-smart pest and disease management) can aid farmers in anticipating and early detecting the spread of pests and diseases (Bouri et al., 2023). Therefore, early detection enables them to provide chemical pesticides and decrease crop losses (Sen & Banerjee, 2023). Consequently, the soil health parameters such as nutrient content (Singh et al., 2023), organic matter content, and soil structure (Ngoc et al. 2023) for soil management such as using cover crops and precision fertilization to enhance soil tolerance and crop yield under extreme climate change can determine by smart soil monitoring systems (Khan et al., 2023).

To mitigate some negative impacts of climate change, such as soil erosion, wind damage, and microclimate fluctuations, innovative planning tools (agroforestry and windbreaks tools) can help farmers in the best designing and planting trees and shrubs as windbreaks into agricultural landscapes to maximize protect ecosystem services and biodiversity (Hoffmeister et al., 2023; Ntawuruhunga et al., 2023).

In addition, To accomplish the goal of conservation agricultural techniques, such as organic fertilization, a forest plantation, reduced tillage, and cover cropping, the integration of smart technologies, such as carbon accounting tools (Whittaker et al., 2013) and carbon markets using blockchain technology can assist farmers to adopt

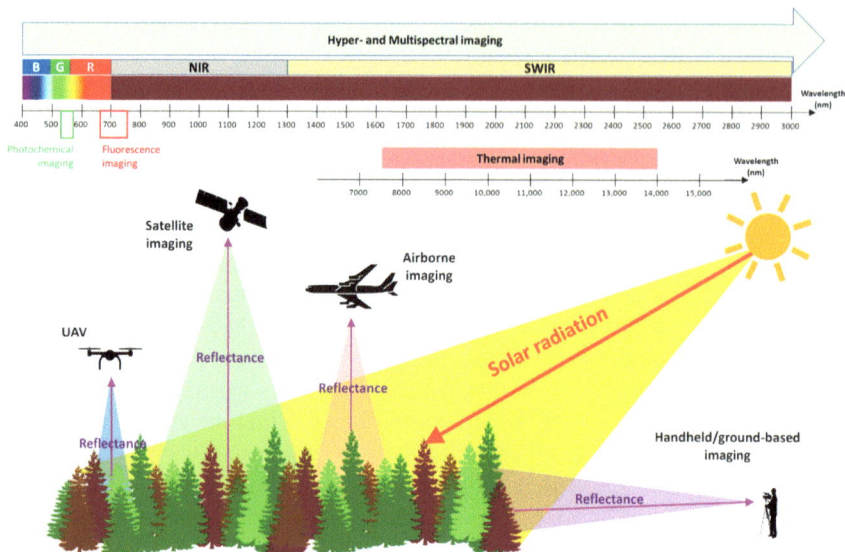

Fig. 8.6 The primary remote sensing platforms and imaging techniques utilized in the identification of forest drought stress (adapted from T.S. Le) (Le et al., 2023)

sustainable practices and commercialize carbon offsets (Mezquita et al., 2023) and definitely can help to enhance soil carbon sequestration and mitigate climate change (Razzaghi, 2021; Razzaghi et al., 2022). Smart techniques by mobile applications, satellite imagery, and weather data, can accurately determine climate-related risks. This method can provide farmers with weather index insurance products that offer financial protection against crop damage under extreme climate change and assist them in recovering from weather-related crises (Hoel-Holt et al., 2023).

To promote collaboration and invention within agricultural systems and information sharing with farmers (Knowledge Sharing and Capacity Building), some smart techniques such as smart agricultural outreach programs, online education initiatives, and peer-to-peer information-sharing platforms can equip farmers with the knowledge needed to apply climate-smart strategies and help them to overcome some disasters that appear suddenly by climate change conditions (Shekmohammed et al., 2023).

As described above, the solutions in the following table (Table 8.1) outline some of the various smart agricultural practices, crop names, impact on the yield of that plant, and references for further information.

Table 8.1 Some of the smart solutions in agriculture in the face of climate change

Category	Solution	Crop	Impact	References
Climate-resilient Crops	Drought-resistant crop	Tomato	No significant difference in yield in comparison with control (without drought stress) NE	(Nahar & Ullah, 2022)
	Heat-tolerant crops	Wheat	Yield ↓	(Mondal et al., 2016)
	Flood-resistant crops	Rice	Survival rate↑	(Septiningsih et al., 2009)
Water management	Drip irrigation	Raspberry and pear	Fruit production and yield↑	(Carroll et al., 2024; Lepaja et al., 2024)
	Rainwater harvesting	Apricot	Controlling the growth and physiological processes of trees	(Feng et al., 2024)
	Smart irrigation systems (IoT)	Saffron	The qualitative and quantitative yield of crop↑	(Iqbal et al., 2024)
Soil management	Conservation tillage	Rice	No-tillage: rice yield↓ Strip tillage: different crop yield by 4.81% ↑	(Dou et al., 2024; Vitali et al., 2024)
	Cover cropping	Cotton	Yield stability in different environmental conditions↑	(Nouri et al., 2020)
	Soil organic matter ↑	Wheat and cotton	Yield↑	(Yang et al., 2024)
Pest and Disease Control	Integrated Pest Management (IPM)	Maize	Yield↑	(Bomami, 2024)

(continued)

Table 8.1 (continued)

Category	Solution	Crop	Impact	References
	Disease-resistant crops	Sunflower	Yield↑	(Habib et al., 2024)
	Bio-pesticides	Eggplants	Yield↑	(Abubakar et al., 2023)
Precision Agriculture	GPS-guided equipment	In general crops	Yield↑	(Schimmelpfennig, 2016)
	Remote sensing	Soybean	Yield prediction models	(Amaral et al., 2024)
	Variable rate technology (VRT)	In general crops	Yield↑	(Papadopoulos et al., 2024)
Agroforestry	Alley cropping	$S.sesban$ + (sorghum + cowpea–barley)	Yield↑	(Kumar et al., 2023)
	Silvopasture	tree + grass + legume with hybrid Napier grass	The protein yield↑ and the quality of forage↑	(Raj et al., 2016)
	Windbreaks	Wheat	Yield↑	(Carberry et al., 2002)
Renewable Energy	Solar-powered irrigation systems	Wheat	Technical efficiency of wheat production ↑	(Ullah et al., 2023)
	Wind energy	In general crops	Indirectly productivity↑	(Chen & Li, 2019)

↑ = Increase, ↓ = Decrease, NE = No effect, LE = Less effect.

8.5   Conclusion

Overall agriculture faces considerable challenges due to climate change, including shifting weather and growing seasons, and water deficiency, that threaten world-wide food security. However, smart solutions create a sense of hope for farmers by empowering them to adapt and thrive in this altering environment. This chapter acts as a gateway to investigate how technology can improve sustainability in agriculture. Comprehending the immediacy of climate change challenges, we lay the groundwork for exploring how smart solutions can pave the way for adaptation to this condition. From precision farming techniques to data-driven decision-making processes, these innovative technologies offer the potential to mitigate climate change effects. Farmers can adapt to climate challenges and enhance crop yields by utilizing smart techniques for future generations. Despite technological accessibility restrictions, knowledge-sharing in field schools can guarantee that the benefits of technology can be available to farmers. This chapter tried to fill the void in previous literature by offering practical consequences and the utility of smart technologies in farming practices to improve the sustainable agricultural sector and build a better future for agriculture in the face of climate change conditions.

References

Abubakar, M., Yadav, D., Koul, B., & Song, M. (2023). Efficacy of eco-friendly bio-pesticides against the whitefly Bemisia tabaci (Gennadius) for sustainable Eggplant Cultivation in Kebbi State. *Nigeria. Agronomy, 13*(12), 3083. https://doi.org/10.3390/agronomy13123083

Ali, A., Hussain, T., Tantashutikun, N., Hussain, N., & Cocetta, G. (2023). Application of smart techniques, internet of things, and data mining for resource use efficient and sustainable crop production. *Agriculture, 13*(2), 397. https://doi.org/10.3390/agriculture13020397

Alipio, M., & Villena, M. L. (2023). Intelligent wearable devices and biosensors for monitoring cattle health conditions: A review and classification. *Smart Health, 27*, 100369. https://doi.org/10.1016/j.smhl.2022.100369

Amaral, L. R., Oldoni, H., Baptista, G. M., Ferreira, G. H., Freitas, R. G., Martins, C. L.,& Santos, A. F. (2024). Remote sensing imagery to predict soybean yield: a case study of vegetation indices contribution. *Precision Agriculture*, 1–19. https://doi.org/10.1007/s11119-024-10174-5

Ammar, A., Iftikhar, Z., Khan, U., Bibi, A., Tahseen, N., Haider, I., & Amjad, I. (2023). Global collaborations in breeding crops for climate resilience. *Biological and Agricultural Sciences Research Journal, 2023*(1), 25–25. https://doi.org/10.54112/basrj.v2023i1.25

Amoak, D., Luginaah, I., & McBean, G. (2022). Climate change, food security, and health: Harnessing Agroecology to build climate-resilient communities. *Sustainability, 14*(21), 13954. https://doi.org/10.3390/su142113954

Andati, P., Majiwa, E., Ngigi, M., Mbeche, R., & Ateka, J. (2023). Effect of climate-smart agriculture technologies on crop yields: Evidence from potato production in Kenya. *Climate Risk Management, 41*, 100539. https://doi.org/10.1016/j.crm.2023.100539

Arif, M., Jan, T., Munir, H., Rasul, F., Riaz, M., Fahad, S., & Amanullah. (2020). Climate-smart agriculture: assessment and adaptation strategies in changing climate. *Global Climate Change and Environmental Policy: Agriculture Perspectives*, 351–377. https://doi.org/10.1007/978-981-13-9570-3_12

Ashraf, H., & Akanbi, M. T. (2023). Sustainable agriculture in the digital age: Crop Management and yield forecasting with IoT, Cloud, and AI. *Tensorgate Journal of Sustainable Technology and Infrastructure for Developing Countries, 6*(1), 64–71.

Bechar, A., & Vigneault, C. (2017). Agricultural robots for field operations. Part 2: Operations and systems. *Biosystems Engineering, 153*, 110–128. https://doi.org/10.1016/j.biosystemseng.2016.11.004

Bell, J. E., Brown, C. L., Conlon, K., Herring, S., Kunkel, K. E., Lawrimore, J., & Uejio, C. (2018). Changes in extreme events and the potential impacts on human health. *Journal of the Air and Waste Management Association, 68*(4), 265-287. https://doi.org/10.1080/10962247.2017.1401017

Beniston, M. (2010). Climate change and its impacts: Growing stress factors for human societies. *International Review of the Red Cross, 92*(879), 557–568. https://doi.org/10.1017/S1816383110000342

Benyam, A. A., Soma, T., & Fraser, E. (2021). Digital agricultural technologies for food loss and waste prevention and reduction: Global trends, adoption opportunities and barriers. *Journal of Cleaner Production, 323*, 129099. https://doi.org/10.1016/j.jclepro.2021.129099

Bomami, D. (2024). Effect of pest management strategies on crop damage and yield in maize in Tanzania. *American Journal of Agriculture, 6*(2), 36–47.

Bouri, M., Arslan, K. S., & Şahin, F. (2023). Climate-smart pest management in sustainable agriculture: Promises and challenges. *Sustainability, 15*(5), 4592. https://doi.org/10.3390/su15054592

Carberry, P., Meinke, H., Poulton, P., Hargreaves, J., Snell, A., & Sudmeyer, R. (2002). Modelling crop growth and yield under the environmental changes induced by windbreaks. 2. Simulation of potential benefits at selected sites in Australia. *Australian Journal of Experimental Agriculture, 42*(6), 887–900. https://doi.org/10.1071/EA02020

Carroll, J. L., Orr, S. T., Benedict, C. A., DeVetter, L. W., & Bryla, D. R. (2024). Feasibility of using pulse drip irrigation for increasing growth, yield, and water productivity of red raspberry. *HortScience, 59*(3), 332–339. https://doi.org/10.21273/HORTSCI17467-23

Çetin, Ö., Yıldırım, M., Akıncı, C., & Yarosh, A. (2022). Critical threshold temperatures and rainfall indeclining grain yield of durum wheat (Triticum durum Desf.) during crop development stages. *Romanian Agricultural Research, 39*, 247–257, DII 2067-5720 RAR 2022-48

Brooks, S., & Loevinsohn, M. (2011). Shaping agricultural innovation systems responsive to food insecurity and climate change. *Natural Resources Forum.* https://doi.org/10.1111/j.1477-8947.2011.01396.x

Chandrappa, V. Y., Ray, B., Ashwatha, N., & Shrestha, P. (2023). Spatiotemporal modeling to predict soil moisture for sustainable smart irrigation. *Internet of Things, 21*, 100671. https://doi.org/10.1016/j.iot.2022.100671

Chen, T., & Li, Q. (2019). Wind energy and agricultural production: Evidence from farm-level data. *Available at SSRN 4854865.*

Chergui, N., & Kechadi, M. T. (2022). Data analytics for crop management: A big data view. *Journal of Big Data, 9*(1), 1–37. https://doi.org/10.1186/s40537-022-00668-2

Chole, A., Jadhav, A., & Shinde, V. (2021). Vertical farming: Controlled environment agriculture. *Just Agric, 1*, 249–256.

Control, C. F. D., & Prevention. (2002). Heat-related deaths--four states, July-August 2001, and United States, 1979–1999. *MMWR: Morbidity and mortality weekly report, 51*(26), 567–570.

Cowie, J. (2012). Climate change: Biological and human aspects. *Cambridge University Press.* https://doi.org/10.1017/CBO9781139087735

Darshna, S., Sangavi, T., Mohan, S., Soundharya, A., & Desikan, S. (2015). Smart irrigation system. *IOSR Journal of Electronics and Communication Engineering (IOSR-JECE), 10*(3), 32–36.

Dou, S., Wang, Z., Tong, J., Shang, Z., Deng, A., Song, Z., & Zhang, W. (2024). Strip tillage promotes crop yield in comparison with no tillage based on a meta-analysis. *Soil and Tillage Research, 240*, 106085. https://doi.org/10.1016/j.still.2024.106085

Drebenstedt, I., Marhan, S., Poll, C., Kandeler, E., & Högy, P. (2023). Annual cumulative ambient precipitation determines the effects of climate change on biomass and yield of three important field crops. *Field Crops Research, 290*, 108766. https://doi.org/10.1016/j.fcr.2022.108766

Engler, N., & Krarti, M. (2021). Review of energy efficiency in controlled environment agriculture. *Renewable and Sustainable Energy Reviews, 141*, 110786. https://doi.org/10.1016/j.rser.2021.110786

Fahad, S., & Wang, J. (2018). Farmers' risk perception, vulnerability, and adaptation to climate change in rural Pakistan. *Land Use Policy, 79*, 301–309. https://doi.org/10.1016/j.landusepol.2018.08.018

Fahad, S., & Wang, J. (2020). Climate change, vulnerability, and its impacts in rural Pakistan: A review. *Environmental Science and Pollution Research, 27*, 1334–1338. https://doi.org/10.1007/s11356-019-06878-1

Farooq, A., Farooq, N., Akbar, H., Hassan, Z. U., & Gheewala, S. H. (2023). A critical review of climate change impact at a global scale on cereal crop production. *Agronomy, 13*(1), 162. https://doi.org/10.3390/agronomy13010162

Feng, N., Huang, Y., Tian, J., Wang, Y., Ma, Y., & Zhang, W. (2024). Effects of a rainwater harvesting system on the soil water, heat and growth of apricot in rain-fed orchards on the Loess Plateau. *Scientific Reports, 14*(1), 9269. https://doi.org/10.1038/s41598-024-58667-7

Fountas, S., Mylonas, N., Malounas, I., Rodias, E., Hellmann Santos, C., & Pekkeriet, E. (2020). Agricultural robotics for field operations. *Sensors, 20*(9), 2672. https://doi.org/10.3390/s20092672

Gentle, P., & Maraseni, T. N. (2012). Climate change, poverty and livelihoods: Adaptation practices by rural mountain communities in Nepal. *Environmental Science and Policy, 21*, 24–34. https://doi.org/10.1016/j.envsci.2012.03.007

Gregory, P. J., Ingram, J. S., & Brklacich, M. (2005). Climate change and food security. *Philosophical Transactions of the Royal Society B: Biological Sciences, 360*(1463), 2139–2148. https://doi.org/10.1098/rstb.2005.1745

Habib, S., Qamar, R., Hassan, E. U., Hussain, F., Anwer, M., Mustafa, S. B., & Khan, M. E. (2024). Orisun-701, A new high yielding and disease resistant sunflower hybrid released in Pakistan. *Pakistan Journal of Phytopathology, 36*(1), 175–184. https://doi.org/10.33866/phytopathol.036.01.1119

Habibullah, M. S., Din, B. H., Tan, S.-H., & Zahid, H. (2022). Impact of climate change on biodiversity loss: Global evidence. *Environmental Science and Pollution Research, 29*(1), 1073–1086. https://doi.org/10.1007/s11356-021-15702-8

Halachmi, I., Guarino, M., Bewley, J., & Pastell, M. (2019). Smart animal agriculture: Application of real-time sensors to improve animal well-being and production. *Annual Review of Animal Biosciences, 7*, 403–425. https://doi.org/10.1146/annurev-animal-020518-114851

Hoel-Holt, A., Skjeflo, S. W., & Vennemo, H. (2023). Climate insurance in developing countries. *CICERO Report*.

Hoffmeister, S., Bohn Reckziegel, R., du Toit, B., Hassler, S. K., Kestel, F., Maier, R., & Zehe, E. (2023). Hydrological and pedological effects of combining Italian alder and blackberries in an agroforestry windbreak system in South Africa. *Hydrology and Earth System Sciences Discussions, 2023*, 1–24. https://doi.org/10.5194/hess-2023-217

Hung, H.-C., Yang, C.-Y., Chien, C.-Y., & Liu, Y.-C. (2016). Building resilience: Mainstreaming community participation into integrated assessment of resilience to climatic hazards in metropolitan land use management. *Land Use Policy, 50*, 48–58. https://doi.org/10.1016/j.landusepol.2015.08.029

Inoue, Y. (2020). Satellite-and drone-based remote sensing of crops and soils for smart farming–a review. *Soil Science and Plant Nutrition, 66*(6), 798–810. https://doi.org/10.1080/00380768.2020.1738899

Iqbal, A., Taqvi, S. A. A., Asim, J., Muneeb, Q., Mahajan, G., Sambyal, R., & Anand, S. (2024). Design of an IOT-based saffron crop irrigation system. *Industrial Crops and Products, 212*, 118350. https://doi.org/10.1016/j.indcrop.2024.118350

Javaid, M., Haleem, A., Singh, R. P., & Suman, R. (2022). Enhancing smart farming through the applications of agriculture 4.0 technologies. *International Journal of Intelligent Networks, 3*, 150–164. https://doi.org/10.1016/j.ijin.2022.09.004

Khan, A., Hassan, M., & Shahriyar, A. K. (2023). Optimizing onion crop management: A smart agriculture framework with IOT sensors and cloud technology. *Applied Research in Artificial Intelligence and Cloud Computing, 6*(1), 49–67.

Khan, J. I., Khan, J., Ali, F., Ullah, F., Bacha, J., & Lee, S. (2022). Artificial intelligence and internet of things (AI-IoT) technologies in response to COVID-19 pandemic: A systematic review. *Ieee Access, 10*, 62613–62660. https://doi.org/10.1109/ACCESS.2022.3181605

Khan, M. D., Thi Vu, H. H., Lai, Q. T., & Ahn, J. W. (2019). Aggravation of human diseases and climate change nexus. *International journal of environmental research and public health, 16*(15), 2799. https://doi.org/10.3390/ijerph16152799

Kiem, A. S., & Austin, E. K. (2013). Drought and the future of rural communities: Opportunities and challenges for climate change adaptation in regional Victoria. *Australia. Global Environmental Change, 23*(5), 1307–1316. https://doi.org/10.1016/j.gloenvcha.2013.06.003

Koomey, J. G., Scott Matthews, H., & Williams, E. (2013). Smart everything: Will intelligent systems reduce resource use? *Annual Review of Environment and Resources, 38*, 311–343. https://doi.org/10.1146/annurev-environ-021512-110549

Kshetri, N. (2021). Blockchain and supply chain management. *Elsevier*. https://doi.org/10.1016/B978-0-323-89934-5.00009-X

Kumar, A., Nagar, S., & Anand, S. (2021). Climate change and existential threats. In *Global climate change* (pp. 1–31). Elsevier. https://doi.org/10.1016/B978-0-12-822928-6.00005-8

Kumar, S., Kumar, T. K., Prasad, M., Singh, J., Choudhary, M., Dixit, A. K., & Ghosh, P. K. (2023). Alley cropping system in degraded land of central India: Evaluation of crop performance, economic benefit, and soil nutrients availability. *International Journal of Plant Production, 17*(1), 81–93. https://doi.org/10.1007/s42106-022-00228-x

Lal, D., Chauhan, C., Joshi, A., Deo, I., & Singh, S. (2024). Smart Breeding for Climate-Resilient Agriculture. In: *Smart Breeding* (pp. 155–167). Apple Academic Press. https://doi.org/10.1201/9781003361862-6

Lavorato, D., & Piedepalumbo, P. (2023). How smart technologies affect the decision-making and control system of food and beverage companies—A case study. *Sustainability, 15*(5), 4292. https://doi.org/10.3390/su15054292

Le, T. S., Harper, R., & Dell, B. (2023). Application of remote sensing in detecting and monitoring water stress in forests. *Remote Sensing, 15*(13), 3360. https://doi.org/10.3390/rs15133360

Lepaja, L., Lepaja, K., Kullaj, E., & Balaj, N. (2024). The influence of drip irrigation on water efficiency in pear cultivation. *Journal of Ecological Engineering, 25*(7), 241–245. https://doi.org/10.12911/22998993/188579

Lowenberg-DeBoer, J. (2015). The precision agriculture revolution. *Foreign Affairs, 94*, 105.

Mac Kinnon, M. A., Brouwer, J., & Samuelsen, S. (2018). The role of natural gas and its infrastructure in mitigating greenhouse gas emissions, improving regional air quality, and renewable resource integration. *Progress in Energy and Combustion Science, 64*, 62–92. https://doi.org/10.1016/j.pecs.2017.10.002

Magiri, R. B., Sagero, P., Danmaigoro, A., Rashid, R., Mocevakaca, W., Singh, S., & Iji, P. A. (2023). Impact of climate change on the dairy production in fiji and the pacific island countries and territories: An insight for adaptation planning.

Maheswari, B. U., Imambi, S. S., Hasan, D., Meenakshi, S., Pratheep, V., & Boopathi, S. (2023). Internet of things and machine learning-integrated smart robotics. In: *Global Perspectives on Robotics and Autonomous Systems: Development and Applications* (pp. 240–258). IGI Global. https://doi.org/10.4018/978-1-6684-7791-5.ch010

Majeed, Y., Khan, M. U., Waseem, M., Zahid, U., Mahmood, F., Majeed, F., & Raza, A. (2023). Renewable energy as alternative source for energy management in agriculture. *Energy Reports, 10*, 344–359. https://doi.org/10.1016/j.egyr.2023.06.032an

Malhi, Y., Franklin, J., Seddon, N., Solan, M., Turner, M. G., Field, C. B., & Knowlton, N. (2020). Climate change and ecosystems. *Philosophical Transactions: Biological Sciences, 375*(1794), 1–8. https://doi.org/10.1098/rstb.2019.0104

Market, I. V. (2018). Global agricultural robots: market size, status and forecast to 2025. *Boonton, NJ, EEUU*.

Meemken, E.-M., & Bellemare, M. F. (2020). Smallholder farmers and contract farming in developing countries. *Proceedings of the National Academy of Sciences, 117*(1), 259–264. https://doi.org/10.1073/pnas.1909501116

Mezquita, Y., Álvarez, C., Valdeolmillos, D., & Prieto, J. (2023). Blockchain-based platforms for carbon offsetting: a survey of existing approaches and their potential to promote carbon farming for smallholders. *International Congress on Blockchain and Applications.* https://doi.org/10.1007/978-3-031-45155-3_47

Mondal, S., Singh, R., Mason, E., Huerta-Espino, J., Autrique, E., & Joshi, A. (2016). Grain yield, adaptation and progress in breeding for early-maturing and heat-tolerant wheat lines in South Asia. *Field Crops Research, 192*, 78–85. https://doi.org/10.1016/j.fcr.2016.04.017

Morais, R., Valente, A., & Serôdio, C. (2005). A wireless sensor network for smart irrigation and environmental monitoring: A position article. 5th European federation for information technology in agriculture, food and environement and 3rd world congress on computers in agriculture and natural resources (EFITA/WCCA).

Muluneh, M. G. (2021). Impact of climate change on biodiversity and food security: A global perspective—a review article. *Agriculture and Food Security, 10*(1), 1–25. https://doi.org/10.1186/s40066-021-00318-5

Mushunje, S. O. O. A. Heterogeneous treatment effect estimation of participation in collective actions and adoption of climate-smart farming technologies in South–West Nigeria.

Mutengwa, C. S., Mnkeni, P., & Kondwakwenda, A. (2023). Climate-smart agriculture and food security in southern Africa: a review of the vulnerability of smallholder agriculture and food security to climate change. *Sustainability, 15*(4), 2882. https://doi.org/10.3390/su15042882

Nahar, K., & Ullah, S. (2022). Climate smart agriculture with drought resistant tomato (Solanum lycopersicum) cultivars under subtropical climate. *International Journal Environmental Climate Change, 12*(12), 138–147. https://doi.org/10.9734/ijecc/2022/v12i121448

Nam, E., & Jin, T. (2021). Mitigating carbon emissions by energy transition, energy efficiency, and electrification: Difference between regulation indicators and empirical data. *Journal of Cleaner Production, 300*, 126962. https://doi.org/10.1016/j.jclepro.2021.126962

Nam, L. P., Van Song, N., Quilloy, A. J. A., Rañola, R. F., Camacho Jr, J. V., Camacho, L. D., & Eluriagac, L. M. T. (2023). Assessment of impacts of adaptation measures on rice farm economic performance in response to climate change: Case study in Vietnam. *Environment, Development and Sustainability*, 1–29. https://doi.org/10.1007/s10668-023-04301-x

Neethirajan, S. (2017). Recent advances in wearable sensors for animal health management. *Sensing and Bio-Sensing Research, 12*, 15–29. https://doi.org/10.1016/j.sbsr.2016.11.004

Newbold, T., Oppenheimer, P., Etard, A., & Williams, J. J. (2020). Tropical and Mediterranean biodiversity is disproportionately sensitive to land-use and climate change. *Nature Ecology & Evolution, 4*(12), 1630–1638. https://doi.org/10.1038/s41559-020-01303-0

Ngoc, T. T. H., Khanh, P. T., & Pramanik, S. (2023). Smart agriculture using a soil monitoring system. In: *Handbook of Research on AI-Equipped IoT Applications in High-Tech Agriculture* (pp. 200–220). IGI Global. https://doi.org/10.4018/978-1-6684-9231-4.ch011

Nguyen, T. T., Grote, U., Neubacher, F., Do, M. H., & Paudel, G. P. (2023). Security risks from climate change and environmental degradation: Implications for sustainable land use transformation in the Global South. *Current Opinion in Environmental Sustainability, 63*, 101322. https://doi.org/10.1016/j.cosust.2023.101322

Nouri, A., Lee, J., Yoder, D. C., Jagadamma, S., Walker, F. R., Yin, X., & Arelli, P. (2020). Management duration controls the synergistic effect of tillage, cover crop, and nitrogen rate on cotton yield and yield stability. *Agriculture, Ecosystems and Environment, 301*, 107007. https://doi.org/10.1016/j.agee.2020.107007

Ntawuruhunga, D., Ngowi, E. E., Mangi, H. O., Salanga, R. J., & Shikuku, K. M. (2023). Climate-smart agroforestry systems and practices: A systematic review of what works, what doesn't work, and why. *Forest Policy and Economics, 150*, 102937. https://doi.org/10.1016/j.forpol.2023.102937

Papadopoulos, G., Arduini, S., Uyar, H., Psiroukis, V., Kasimati, A., & Fountas, S. (2024). Economic and environmental benefits of digital agricultural technologies in crop production: A review. *Smart Agricultural Technology*, 100441. https://doi.org/10.1016/j.atech.2024.100441

Pascaris, A. S., Schelly, C., Burnham, L., & Pearce, J. M. (2021). Integrating solar energy with agriculture: Industry perspectives on the market, community, and socio-political dimensions of agrivoltaics. *Energy Research and Social Science, 75*, 102023. https://doi.org/10.1016/j.erss.2021.102023

Pathak, T. B., Maskey, M. L., Dahlberg, J. A., Kearns, F., Bali, K. M., & Zaccaria, D. (2018). Climate change trends and impacts on California agriculture: A detailed review. *Agronomy, 8*(3), 25. https://doi.org/10.3390/agronomy8030025

Prasad, P., Bhatnagar, N., Bhandari, V., Jacob, G., Narayan, K., Echeverría, R., & Compton, J. (2023). Patterns of investment in agricultural research and innovation for the Global South, with a focus on sustainable agricultural intensification. *Frontiers in Sustainable Food Systems, 7*, 1108949. https://doi.org/10.3389/fsufs.2023.1108949

Raj, A. K., Kunhamu, T., Jamaludheen, V., & Kiroshima, S. (2016). Forage yield and nutritive value of intensive silvopasture systems under cut and carry scheme in humid tropics of Kerala, India. *Indian Journal of Agroforestry, 18*(1), 47–52.

Rajak, P., Ganguly, A., Adhikary, S., & Bhattacharya, S. (2023). Internet of Things and smart sensors in agriculture: Scopes and challenges. *Journal of Agriculture and Food Research, 14*, 100776. https://doi.org/10.1016/j.jafr.2023.100776

Razzaghi, S. (2021). Effects of cover crops on greenhouse gas emissions. In: *Cover Crops and Sustainable Agriculture* (pp. 280–298). CRC Press. https://doi.org/10.1201/9781003187301-16

Razzaghi, S., Islam, K. R., & Ahmed, I. A. M. (2022). Deforestation impacts soil organic carbon and nitrogen pools and carbon lability under Mediterranean climates. *Journal of Soils and Sediments, 22*(9), 2381–2391. https://doi.org/10.1007/s11368-022-03244-6

Rogers, D., & Randolph, S. (2006). Climate change and vector-borne diseases. *Advances in Parasitology, 62*, 345–381. https://doi.org/10.1016/S0065-308X(05)62010-6

Sadowski, J. (2020). *Too smart: How digital capitalism is extracting data, controlling our lives, and taking over the world*. mit Press. https://doi.org/10.7551/mitpress/12240.001.0001

Saiz-Rubio, V., & Rovira-Más, F. (2020). From smart farming towards agriculture 5.0: A review on crop data management. *Agronomy, 10*(2), 207. https://doi.org/10.3390/agronomy10020207

Salah, A., Oumarou, A., Ismaila, M. H., & Amadou, O. (2023). *Development of IOT based smart irrigation system* department of mechanical and production engineering (MPE), Islamic

Salvini, G., Van Paassen, A., Ligtenberg, A., Carrero, G., & Bregt, A. (2016). A role-playing game as a tool to facilitate social learning and collective action towards Climate Smart Agriculture: Lessons learned from Apuí, Brazil. *Environmental Science and Policy, 63*, 113–121. https://doi.org/10.1016/j.envsci.2016.05.016

Sarker, I. H. (2021). Data science and analytics: An overview from data-driven smart computing, decision-making and applications perspective. *SN Computer Science, 2*(5), 377. https://doi.org/10.1007/s42979-021-00765-8

Sasikala, S., & Bharatula, S. D. (2024). IoT-Based smart irrigation system based adaptive radial deep neural network (ARDNN) algorithm applicable for various agricultural production. *International Journal of Intelligent Systems and Applications in Engineering, 12*(13s), 351–366–351–366.

Sathaye, J., Shukla, P., & Ravindranath, N. (2006). Climate change, sustainable development and India: Global and national concerns. *Current science*, 314–325.

Schimmelpfennig, D. (2016). Farm profits and adoption of precision agriculture.

Sen, K., & Banerjee, A. (2023). Insect pest management under climate change scenario. *ENTOMOLOGY, 133*, 35.

Septiningsih, E. M., Pamplona, A. M., Sanchez, D. L., Neeraja, C. N., Vergara, G. V., Heuer, S.,…Mackill, D. J. (2009). Development of submergence-tolerant rice cultivars: The Sub1 locus and beyond. *Annals of Botany, 103*(2), 151–160. https://doi.org/10.1093/aob/mcn206

Shekmohammed, S., Hany, U., & Lemma, S. (2023). Review of farmers field school approach for facilitation of climate smart agriculture. *International Journal Agriculture Veterinarary Science, 5*(1), 9–17. https://doi.org/10.34104/ijavs.023.09017

Silvestre, B. S., & Țîrcă, D. M. (2019). Innovations for sustainable development: Moving toward a sustainable future. *Journal of Cleaner Production, 208*, 325–332. https://doi.org/10.1016/j.jclepro.2018.09.244

Singh, H., Halder, N., Singh, B., Singh, J., Sharma, S., & Shacham-Diamand, Y. (2023). Smart farming revolution: Portable and real-time soil nitrogen and phosphorus monitoring for sustainable agriculture. *Sensors, 23*(13), 5914. https://doi.org/10.3390/s23135914

Singh, R., Machanuru, R., Singh, B., & Shrivastava, M. (2021). Climate-resilient agriculture: enhance resilience toward climate change. In: *Global climate change* (pp. 45–61). Elsevier. https://doi.org/10.1016/B978-0-12-822928-6.00016-2

Sinha, B. B., & Dhanalakshmi, R. (2022). Recent advancements and challenges of Internet of Things in smart agriculture: A survey. *Future Generation Computer Systems, 126*, 169–184. https://doi.org/10.1016/j.future.2021.08.006

Smith, H. E., Sallu, S. M., Whitfield, S., Gaworek-Michalczenia, M. F., Recha, J. W., Sayula, G. J., & Mziray, S. (2021). Innovation systems and affordances in climate-smart agriculture. *Journal of Rural Studies, 87*, 199–212. https://doi.org/10.1016/j.jrurstud.2021.09.001

Stankovic, M., Neftenov, N., & Gupta, R. (2022). Use of digital tools in fighting climate change: A review of best practices. Retrieved from https://bit.ly/3Gxodt6 (Access date: 21 October 2022).

Thornton, P., Nelson, G., Mayberry, D., & Herrero, M. (2021). Increases in extreme heat stress in domesticated livestock species during the twenty-first century. *Global Change Biology, 27*(22), 5762–5772. https://doi.org/10.1111/gcb.15825

Ullah, I., Khan, N., Dai, Y., & Hamza, A. (2023). Does solar-powered irrigation system usage increase the technical efficiency of crop production? *New Insights from Rural Areas. Energies, 16*(18), 6641. https://doi.org/10.3390/en16186641

Vatistas, C., Avgoustaki, D. D., & Bartzanas, T. (2022). A systematic literature review on controlled-environment agriculture: How vertical farms and greenhouses can influence the sustainability and footprint of urban microclimate with local food production. *Atmosphere, 13*(8), 1258. https://doi.org/10.3390/atmos13081258

Vijai, C., Worakamol, W., & Elayaraja, M. (2023). Climate change and its impact on agriculture. https://doi.org/10.25303/1104ijasvm0108

Vitali, A., Moretti, B., Lerda, C., Said-Pullicino, D., Celi, L., & Romani, M.,…Vidotto, F. (2024). Conservation tillage in temperate rice cropping systems: Crop production and soil fertility. *Field Crops Research, 308*, 109276. https://doi.org/10.1016/j.fcr.2024.109276

Wade, K., & Jennings, M. (2016). The impact of climate change on the global economy. *Schroders Talking Point.*

Whittaker, C., McManus, M. C., & Smith, P. (2013). A comparison of carbon accounting tools for arable crops in the United Kingdom. *Environmental Modelling and Software, 46*, 228–239. https://doi.org/10.1016/j.envsoft.2013.03.015

Xiong, W., Li, Y., Pfister, S., Zhang, W., Wang, C., & Wang, P. (2020). Improving water ecosystem sustainability of urban water system by management strategies optimization. *Journal of Environmental Management, 254*, 109766. https://doi.org/10.1016/j.jenvman.2019.109766

Yadav, N., & Sidana, N. (2023). Precision Agriculture Technologies: Analysing the use of advanced technologies, such as drones, sensors, and GPS, In: Precision agriculture for optimizing resource management, Crop Monitoring, and Yield Prediction. *Journal of Advanced Zoology, 44.*

Yang, C., Wang, X., Li, J., Zhang, G., Shu, H., Hu, W., & Guo, Z. (2024). Straw return increases crop production by improving soil organic carbon sequestration and soil aggregation in a long-term wheat–cotton cropping system. *Journal of Integrative Agriculture, 23*(2), 669–679. https://doi.org/10.1016/j.jia.2023.06.009

Zhang, Z., Pan, S.-Y., Li, H., Cai, J., Olabi, A. G., Anthony, E. J., & Manovic, V. (2020). Recent advances in carbon dioxide utilization. *Renewable and Sustainable Energy Reviews, 125*, 109799. https://doi.org/10.1016/j.rser.2020.109799

Chapter 9
Conservation Techniques in Agriculture Under Climate Change

Urfan Taghiyev[ID] **and Ulker Yusubova**[ID]

Abstract Soils, which are the main factor in crop and animal production, are a living and heterogeneous natural resource and face many negative impacts during conventional agricultural activities. This chapter presents an analysis of the results of research on the tillage system. When conducting research, our main goal was to study the negative impact on the soil during the use of agricultural aggregates, as well as the impact of resource-saving technologies on increasing soil fertility and naturally increasing both the productivity of agricultural machines and increasing yields. Conservation agriculture (CA) is a new resource management system compared with the traditional form, biological activity and physical structure are preserved. The CA also provides for the use of resource-saving technologies and organic seed material, as a result of which productivity and yield increase, as well as the number of passes through the field and environmental pollution decrease. In this chapter, the main topics such as soil vitality, minimum and/or zero tillage, weed control, cultivation techniques and economic rationale, which constitute the main function of conservation agriculture, are covered.

Keywords Agriculture · Conservation agriculture · Resource saving · Technology use

9.1 Introduction

Agriculture is based on three basic factors: seed, soil and climate. For agriculture, i.e. crop and animal production, the direct use of soil, which is a natural resource, is mandatory. In order to achieve this, the soils must be cultivated (plowed) properly. If do not properly manage or use soils, it will be degraded in a short time and the quality and health of soils will decrease. Therefore, conservation agriculture (CA) has become extremely important for climate change and sustainable agriculture. Thus,

U. Taghiyev (✉) · U. Yusubova
Azerbaijan State Agricultural University, Ganja, Azerbaijan
e-mail: urfan.tagiyev@adau.edu.az

© The Author(s), under exclusive license to Springer Nature Switzerland AG 2024   173
Ö. Çetin (ed.), *Agriculture and Water Management Under Climate Change*,
SpringerBriefs in Earth System Sciences, https://doi.org/10.1007/978-3-031-74307-8_9

CA is a top priority to prevent degradation of soils and agricultural lands (Çetin et al., 2018).

About a third of the planet's soils are degraded. In many countries, the intensification of crop production has led to soil depletion to such an extent that future production in these areas is at risk. The reduction of agricultural land areas on the globe, the aggravation of food problems and the need to ensure the country's food sovereignty have led in the last decade to attract attention to the tasks of solving them. Healthy soils are the key to building sustainable crop production systems that are resilient to the effects of climate change. This contains a diverse community of organisms that help fight plant diseases, insect populations and weeds. The recycle nutrients improve soil structure, having a positive effect on water retention, nutrient retention and organic carbon conservation (Derpsch, 2003; Garbach et al., 2014).

Conventional agriculture in general has many negative aspects, such as soil erosion, reduction of soil organic matter, water loss, physical degradation of the soil and increased fuel use. Conservation agriculture, on the other hand, helps both to avoid these negative aspects and to improve biodiversity in natural and agricultural ecosystems.

One of the main types of land protection is soil protection and resource-saving agriculture, which is 20–50% less labour-intensive and contributes to reducing greenhouse gas emissions by reducing energy costs and increasing the efficiency of nutrient use. Soil-protective and resource-saving agriculture provides a number of advantages at the global, regional, local and farm levels. It provides also a truly sustainable production system, not only preserving, but also multiplying natural resources and increasing the diversity of soil biota, fauna and flora (including wild ones) in agricultural production systems without damage and risk of reducing high yields. Thus, the main objective of conservation agriculture is a farming system that promotes minimal soil degradation or no-till farming or proper tillage to maintain a permanent soil cover and diversification of plant species. In addition, CA is one of the most important practices for agricultural production and sustainable agriculture, such as increased biodiversity, labour savings, healthier soils and increased productivity.

In this chapter, soil as a living system, conservation agriculture and its adoption, soil tillage techniques and their technical and economic aspects are discussed.

9.2   Definition and Main Principles of Conservation Agriculture

There are many explanations and definitions of the term "conservation agriculture", but these all agree that this is a system that relies on ecosystem management rather than the use of external agricultural resources. This is a system that takes into account the potential harmful effects on the environment and humans of synthetic additives such as synthetic fertilizers and pesticides, veterinary medicines, genetically modified seeds and livestock breeds, preservatives, additives and radiation. All these methods are being replaced in organic agriculture by special methods and practices that preserve and increase soil fertility and prevent the reproduction of pests and the growth of diseases. "CA is an integrated production management system that supports and promotes the health of the agro-ecosystem, including biological diversity, biological cycles and biological activity of the soil. This is a system that focuses on management practices rather than on the use of external agricultural resources, taking into account that specific regional conditions require own systems adapted to the region. All this is accompanied by the application, where possible, of agronomic, biological and mechanical methods, as opposed to the use of synthetic materials to ensure functioning within the system (Montgomery, 2007). Organic farming systems and the products produced by them are not always certified and in this case are called "non-certified organic farming system or product". This does not include agricultural systems that do not use synthetic additives by default (i.e. systems that have no experience in soil reconstruction and deplete the wealth of the land).

CA focused on services. In many countries of the European Union, funds are being created to subsidize organic agriculture that produces environmental products and services, such as combating groundwater pollution or creating a richer and more biologically diverse natural landscape.

There are three main principles and/or components of conservation agriculture which proposal for big- and smallholders (Fig. 9.1) (Komarek & Thierfelder 2021). The first is to till and to disturb the soil as little as possible (Fig. 9.1a); thus, it could be performed minimum mechanical soil and disturbance through direct seed and/or fertilizer placement. The second is to grow different crops (crop rotation) (Fig. 9.1b); this can provide permanent soil organic cover with crop residues and/or cover crops the least 30 percent. The third is to keep the soil permanently covered with vegetation or mulching (Fig. 9.1c). It should be performed species diversification through varied crop sequences and associations involving at least three different crops.

9.2.1 History and Adoption of CA

For many thousands of years, crises have been regional in nature. The largest of those include the crises of agriculture in ancient times (Mesopotamia, Central America), the Middle Ages (Western Europe), and modern history (Russia—XIX century, USA— 30 s of XX century, Asia, Africa—70 s-80 s of XX century) (Dill, 2005). The first food crisis arose already at the beginning of civilization and was caused by the depletion of fishing resources in places of population concentration. It was followed (VII century BC) by the first agricultural revolution associated with the development of agriculture and cattle breeding. But by the beginning of the new era, salinization of soils led to a crisis of primitive irrigation agriculture. Around the tenth century, non-irrigated agriculture was introduced and the lands in the zone of irrigated agriculture on which crops are cultivated without artificial irrigation. That was the second agricultural revolution. For thousands of years, the yield of agricultural crops has been determined

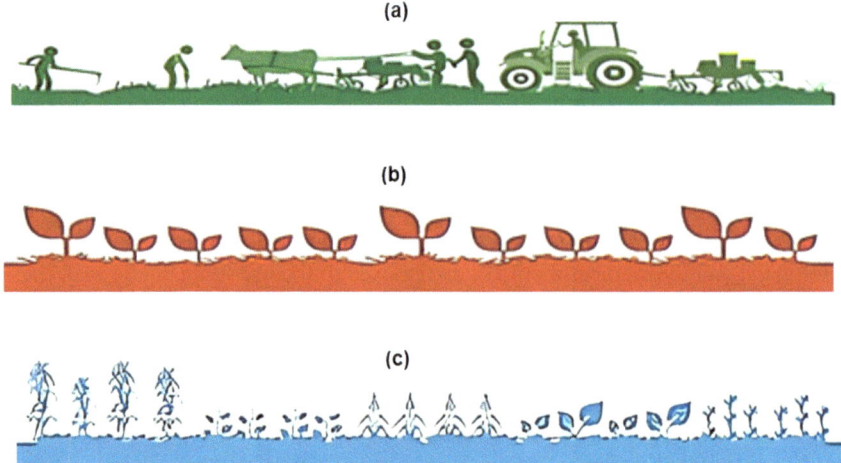

Fig. 9.1 Three principles on conservation agriculture (Komarek and Thierfelder 2021)

by the level of the natural productive potential of soils and the nature of agro-technical techniques for its activation and maintenance.

In different countries, the transition from extensive to intensive agriculture has stretched over many decades. In Azerbaijan, the crisis of extensive agriculture took an exceptionally acute form in the last quarter of the XIX century. The dominance of the three-field system of agriculture, continuous ploughing of land in the steppe zone and the accompanying soil degradation led to catastrophic consequences (Cynthia et al. 1983).

Ecological farming is based on conducting agricultural activities without the use of chemical additives, growth hormones, GMOs and antibiotics. Pesticides and chemical fertilizers are also not used in weed control (Dumanski et al., 2006). Eco-friendly agriculture is a promising area that has a huge benefit for investments and has a number of advantages: it provides good profitability growth; increases the competitiveness of agricultural products; opens a new profitable export channel; gives an additional source of income to the villagers; helps to attract highly qualified specialists to the village; and solves a large number of environmental problems. The yield index in such agriculture is provided by the use of only organic substances: (i) natural fertilizers, (ii) side rates and (iii) probiotics and specialized agricultural technologies. Exclusively biological technologies based on beneficial microorganisms and mechanical techniques are also used to destroy weeds.

9.2.2 Resource Conservation and Agroecology in Agriculture

Achieving sustainable development of the agricultural economy now and in the future requires solving the problem of optimizing resource consumption and resource conservation. The production of agricultural products involves labour (production personnel, etc.), energy (fuels and lubricants of various origins), raw materials (soil, environment, machine and tractor fleet, infrastructure, fertilizers, etc.) and information resources.

The problem of resource conservation should be considered from the perspective of agro-ecological problems of agriculture, crop production systems, machine technologies and machines for complex mechanization of agricultural production, given those are key resources in the production of agricultural products. Non-compliance with agricultural technologies in agriculture has a negative impact on soil fertility and the environment, causing a number of problems.

Improperly carried out ploughing disrupts the structure of the soil and leaves it unprotected from precipitation, contributes to water erosion and pollution of surface waters, reduces the content of organic matter in the soil and the diversity of soil organisms, provokes unnecessary emission of carbon dioxide into the atmosphere, etc. 58.6% of agricultural land is subject to erosion, and more than 1.5 billion tons of fertile layer are lost annually. Water erosion is recorded on 17.8% of agricultural land. According to the rate of soil erosion, the Republic of Azerbaijan occupies one of the first places in the world. The process of "failure" of agricultural lands requires serious agro-technical attention, as it can lead to irreversible consequences, and, as a result, to a sharp reduction in food production opportunities in the future. Land desertification is one of the consequences of erosion. There are about 2 million hectares in Azerbaijan, for which the process of desertification poses a serious threat.

Regularly recurring droughts in the main grain regions of Azerbaijan negatively affect the accumulation of moisture in the soil profile, increase the risk of farming and prevent profitable yields, since moisture deficiency does not fully realize either the genetic potential of varieties or the potential of soil and other resources. The production of 1 ton of grain requires at least 80 tons of moisture. Agriculture in Azerbaijan is one of the main sources of pollution of surface waters, with animal husbandry (runoff) playing a major role. Water discharged from fields carries soil particles, elements of decomposition of pesticides, fertilizers and other organic and inorganic compounds. To mitigate this negative phenomenon, it is necessary to apply a set of measures, the most important of which are the use of agro-landscape farming techniques, the preservation of plant residues on the soil surface and maximum occupancy of the soil by plants.

9.2.3 Soils as the Main Component of CA

Soil, the object of research in many sciences, performs a number of important ecological functions. Conditionally, it can be divided into biogenetics and global. Before starting, it needs to be clarified that scientists from different fields of science have not yet come to an agreement on the definition of the word "soil". If living organisms could occupy all available space, then in less than a year it would be crushed by a mass of microorganisms. One of the main reasons why most organisms are not found everywhere is the lack of a free and habitable material environment. Therefore, most species have only a small percentage of individuals out of a theoretically possible number at any given time (Fawcett, 1987). Nevertheless, living organisms take it "piece of life"—a place where it can live and reproduce relatively peacefully. Therefore, it is no coincidence that in the process of evolution, organisms have mastered the whole shell of the earth, the biosphere, an important part of which is the soil sphere—the pedosphere. It is quite natural that the cycle of development of most plants begins in soils, and in the subsequent stages of the life cycle underground organs (roots) closely interact with the soil. The distribution of roots is uneven both in depth and in geographical altitudes and biosensors: the largest absolute mass of roots is observed in deciduous forests, but if this proceeds from the proportion of root mass in the mass of the whole plant (photomasks), then the steppes will become the leaders (Foresight, 2011). The depth of root penetration depends on the density, distribution of chemical elements and other soil parameters. The soil is abundantly populated with microorganisms: bacteria, archaea, fungi and to a lesser extent algae. Most of these are in the upper layers of the soil, which is not surprising, because there are a lot of delicious organic matters for those. The numbers decrease with depth, but in some areas, such as root passages, there may be more of those than in the upper horizons. It must not forget about seasonal changes (Serraj & Siddique, 2012). In autumn, the number of microorganisms in the soils of the middle zone increases (FAO, 2024). This is due to the intake of a huge amount of food in the form of leaf litter and other plant residues. The soil also serves as a living space for many animals. Almost half of all these types have representatives living in the pedosphere. Invertebrates: flat, round and annelid worms; mollusks; crustaceans; arachnids; insects. Of the vertebrates: amphibians, reptiles, mammals and even some birds (Paarlberg, 2001) Moreover, the soil can act as a completely different environment for organisms, depending on the size. For example, microscopic animals (such as rotifers) essentially remain inhabitants of the aquatic environment. With strong moisture, these float freely in the water; during drought, they accumulate on soil particles and live in so-called water films. For non-microscopic, but still small organisms (mites, medium-sized insects, larvae), life in the soil is similar to living in a moisture-saturated cave. This seems to live not in the soil itself, but in the pore space between the solid particles. For larger animals (earthworms, millipedes, and others), the habitat is the soil as a whole, that is, a loose or dense substrate. Some animals, with the onset of drought, move into the depths, where moisture accumulates and does not evaporate, and in a very humid period, on the contrary, it follows oxygen up. And here immediately recall the appearance of annelid worms on the surface after rain (Friedrich & Kassam., 2009). The soil protects living organisms from hypothermia, overheating, and terrestrial predators, since the temperature and humidity of the air in it are subject to less fluctuations than on the surface. This feature is often used by organisms living in extreme conditions–taiga, desert, etc. This is especially important for animals occupying several environments at once (gophers, voles or hamsters). It gets this food on the surface of the earth and in the soil the hide from predators and bad weather, as well as leave reserves. Due to the supporting function of soils, plants anchored in them by these roots maintain an upright position, are resistant to windfalls and resist gravity.

So, in permafrost areas, the soils are "weak" and can observe a "drunken forest" with bizarre, strongly inclined plants (Friedrich et al., 2012). The supporting function of soils is also manifested in relation to animals. Often, the settlement of soil inhabitants depends on the mechanical properties of the soil. As already mentioned, ground squirrels do not need a dense but also not a loose substrate. An under-studied manifestation of the supporting function is considered to be its effect on the vital activity of terrestrial organisms. The behaviour of animals depends on the conditions of movement. For example, an elk, if necessary, can safely walk through swamps, which cannot be said about other inhabitants of the forest. Due to these properties, most soils serve as an environment in which seeds and other germs (cysts, eggs of invertebrates) are preserved. This is possible due to relatively small differences in temperature and moisture in the soil. The question of the duration of preservation of seeds and germs in the soil is of practical importance. For example, such a strange phenomenon at first glance as overgrowing of cuttings without bringing seeds from the side (Potter et al., 2008).

9.2.3.1 Physico-Chemical Functions

Small (up to 0.25 microns in diameter) colloidal soil particles adsorb gases, liquids, and other molecules. The more such small particles there are, the stronger the absorption, which makes it possible to retain nutrients in the soil that would otherwise be washed out. At the same time, substances can remain accessible to plants, or, conversely, they can be immobilized. There are different ways to optimize this function: liming acidic soils and plastering saline soils, applying organic fertilizers, adding clay to sand fractions, etc. Not only useful elements are retained in soils, but also toxic ones, such as heavy metals. Mercury trapped on the soil surface is washed out very slowly (fractions of a percent per year). As a result of industrial pollution of the atmosphere by aerosols, a lot of dust with toxic substances settles on the soil surface. Therefore, it is necessary to take this function into account when designing factories, landfills, pipelines, etc.

9.2.3.2 The Biological Clock

Many properties of the soil change periodically: there are special thermal, water, salt and food regimes in it. Thus, it has been shown that the leading factor in triggering root growth is soil temperature. A striking example is the acceleration of seasonal plant development during the rainy season in arid regions. The influence of the annual dynamics of the nutritional regime of soils on seasonal changes is noticeable in the fluctuation in the number of microorganisms during the period of abundant leaf fall. Organics are becoming more abundant; organisms are better provided with food and actively reproduce.

9.2.3.3   Transformation of Substances and Energy.

The soil transforms substances falling into its sphere (e.g. rocks), as a result of which favourable conditions for the life of organisms are created. For example, available forms of elements necessary for plant nutrition accumulate in the upper horizons. Or minerals are destroyed by the action of water, acid and vital activity of organisms. An important result of this transformation is the release during decomposition of organic residues of energy accumulated during photosynthesis. This energy is released not only in thermal, but also in chemical form (Miguel et al., 2011).

9.3   The Components of CA

9.3.1   Minimal Soil Disturbance

One of the types of nature-preserving agriculture is minimal tillage. Minimal tillage is a technology that can reduce energy consumption by reducing the number of soil treatments and the depth of tillage. The need to minimize land treatments is determined by the following factors: a decrease in soil productivity and fertility after the operation of heavy aggregates; a decrease in energy consumption by reducing the depth of loosening and the number of such treatments; the possibility of replacing mechanical treatment with the use of herbicides; the need to save time and resources; the emergence of the opportunity to use new agricultural machinery with modern aggregates, combined mechanisms. The main directions of minimal processing: (i) The treatment is combined with the use of herbicides, (ii) the widespread use of herbicides on such soils reduces the number of row-to-row treatments of row crops and (iii) deep processing is replaced by equally effective flat-cutting or surface treatment.

Preference is given to wide-reach units in combination with working mechanisms that ensure high-quality processing in one pass. Several different technological operations are combined into one workflow. It is increasingly practiced to minimize the cost of tillage by combining various fieldworks using various aggregates. However, the implementation of minimal processing requires compliance with certain conditions: (i) Optimal soil density must be maintained or formed for each crop, (ii) for row crops, it is 1.0–1.2 g cm^{-3} and 1.2–1.3 g cm^{-3} for cereals, (iii) the total porosity of the soil should be maintained within 50–55%.

The water permeability of the soil should be ensured at a level of no more than 60 mm/h. The moisture capacity of the soil should be maintained within 30–33%. The thickness of the arable layer should be more than 20–22 cm. The number of harmful organisms in the local agrophytocenosis should be below the threshold of harmfulness. As a rule, well-structured chernozems, chestnut and dark grey forest soils, and any light soil composition are characterized by agrophysical properties favourable for plant organisms. Therefore, such types of soils in principle do not need intensive mechanical treatment.

Disadvantages of minimal tillage: (i) The sanitary condition is deteriorating, and the contamination of crops is noticeably increasing, (ii) the incidence of pests, parasites and diseases on agricultural plants is increasing, (iii) the rate of humus mineralization decreases, and nitrogen intake into the soil decreases, which requires more intensive application of nitrogen fertilizers, and (iv) this is especially true for fields that had stubble precursors.

9.3.2 Weed Control

One of the problematic factors in crop production is weed control. To date, two of those have been used: chemical and biological. The process of environmental agriculture includes, as the main element of preserving biological activity, a biological method, which is included as one of the main elements of weed control. Comprehensive Weed Control Measures (WCM) are aimed at destroying and reducing the number of weeds in the fields and involve a combination of different techniques. The oldest way to solve the problem is mechanical weed control, namely this is manual removal. However, this approach requires a lot of effort and takes a lot of time, especially when the territories are vast and there are few workers. Weed control with herbicides is currently very popular among farmers, but within the framework of sustainable agriculture and organic agriculture, such practices should be preferred to weed control without chemicals. The best option would be an integrated approach.

The removal of weeds with the help of chemistry must be minimized since the remains of herbicides in crops pose a danger to both humans and the environment. Moreover, in many cases, plants develop resistance to chemicals. For these reasons, weed control in the fields should include alternative measures and tactics (FAO, 2008). Effective weed control involves a combination of the five methods listed below. Preventive methods of weed control—The main purpose of this control method is to prevent weed seeds from entering the planting material and cultivated areas. How to protect the field: (i) use high-quality seeds without weed plants, (ii) wash the wheels of the equipment regularly, (iii) inspect the six paws of the animals, (iv) check the irrigation water for weeds, (v) use thoroughly rotted organic fertilizers to prevent unwanted seed germination. Cleaning of the combine and its main assembly units is, as it was, one of the main factors in controlling the spread of weed seed material. This method of control consists of creating less favourable conditions for the development of weeds, which includes: the cultivation of highly adaptive and competitive species; seed treatment and selection of large seeds, which are more likely to produce strong and full of energy shoots; proper alternation of crops in crop rotation; introduction of vapours; the use of cover crops; reducing the width of the aisles; sowing at a minimum depth (so cultivated plants can outstrip weeds in growth); planting local crops that adapt more quickly to growing conditions and compete more easily with weeds than imported plants. However, ploughing the field in autumn contributes to the freezing of weed seeds in winter. In many farms, mechanical as well as agrotechnical weed control measures include: ploughing; weeding; mowing; pulling out (manual removal); burning; mulching; covering the aisles (e.g., with straw); the use of robotic welding machines; the use of weed seed destructors; harvesting hay before the formation of seeds.

9.3.3 Cultivation Techniques or Tillage

The method of mechanical tillage is the nature and degree of influence by the working bodies of tillage implements and machines on the profile change, genetic and anthropological heterogeneity of the cultivated soil layer. During the machining process, there are: dump, non-dump, rotary and combined methods. Dump is a type of tillage by the action of working bodies of tillage implements and machines on the soil with full or partial wrapping of the treated layer to change the location of different-quality layers or genetic horizons of the soil in the vertical direction with increased loosening and mixing of the soil, as well as pruning and embedding of terrestrial plant organs and fertilizers into the soil. All types of dump processing are mainly carried out by ploughings of different designs. Fall-free is a type of processing by working bodies of tillage tools and machines on the soil without changing the location of genetic horizons and differentiation of the treated layer by fertility in the vertical direction in order to loosen the soil, prune underground and preserve aboveground plant organs on the surface of the soil horizon. With this method, stubble is preserved on the soil surface. The waste-free method of tillage is carried out by ploughings with removed dumps, chisel ploughings, chisel cultivators, heavy cultivators. Rotary type of mechanical treatment is the effect on the soil by rotating working bodies of tillage implements and machines to eliminate the differentiation of the treated layer by addition and fertility by active crumbling and mixing of soil, plant residues and fertilizers to form a homogeneous soil layer. This type of processing is carried out by four cuts (FAO, 2011a).

Combined processing methods—Various combinations of horizons and soil layers, as well as the timing of dump, dump less and rotary processing methods. The use of one of the types of processing is due to climatic conditions, soil type and degree of cultivation, and the requirements of cultivated crops. Mechanical processing is a single impact on the soil by various types of tillage tools and machines in one way or another in order to carry out one or more technological operations to a certain depth. Depending on the depth of tillage, 4 groups of techniques are distinguished: surface, conventional, deep and ultra-deep tillage.

Rolling—This method of tillage is carried out by rollers, which ensure the crumbling of lumps, lumps, compaction and levelling of the soil surface. It can also be pre-sowing and post-sowing. Pre-sowing rolling is a mandatory treatment technique, especially on peat and light but granulomeres sandy and sandy loam soils. On light soils, post-sowing rolling at the same time as sowing will also have a great effect. Smooth, ring-spur, ring-toothed, etc. rollers are used for rolling.

Disking is a method of tillage that provides crumbling, loosening, partial wrapping and mixing of the soil, crushing of weeds. The disc harrow as a working organ has rotating spherical discs that can be installed at different angles of attack to the direction of movement. With an increase in the angle of attack, crumbling and processing depth increase, weeds are better cut. Harrows with cut-out discs are used on heavy and grainy soils. Stubble peeling is a soil tillage technique that is carried out after harvesting grain crops, providing crumbling, loosening, partial mixing and wrapping of the soil, crushing of underground and sealing of aboveground plant organs, weed seeds, pathogens and pests of cultivated plants with dump or disc huskers. This wraps and loosens the soil to a depth of 6 to 12 cm and cuts horizontally arranged rhizomes well, ploughshares wrap the soil well and prune weeds to a depth of 8–16 cm. Chisel cultivators can be used for peeling stubble.

Cultivation is a type of crumbling, loosening, mixing of soil, pruning of underground organs of weeds. The working organs of cultivators are paws of various designs. Cultivators loosen the soil from 6 to 12 cm. In places prone to wind erosion, cultivators KPSh-5, KPSh-9 and rod cultivators OP-8.5 and OP-12 are used to leave stubble on the soil surface. Alignment or looping is a method of treatment by levelling the surface of loose soil. This type is carried out by cultivators with simultaneous harrowing, combined units such as CD-720 M, AKSh and RVK, heavy spring harrows BSP-15, BSP-21, wooden beams, travois, etc.

Combing is a type of processing that provides a form of changing the surface of the field for better warming and earlier maturation of the soil, performed by working bodies such as a hiller; ridges promote the formation of ridges on the surface of the field, faster warming and maturation of the soil. Furrowing is the cutting of furrows on the surface of the soil by hoppers-furrowers.

Hollowing is the formation of closed recesses of the soil by disk lunk formers to detain melt water and storm water on soils subject to water erosion. Hoeing is a type of row-to-row cultivation with the soil being rolled to the base of the stems of row crops by the working bodies of cultivators of hoppers. The bouquet provides thinning of beet seedlings with a given size of cutouts and bouquets, crumbling, loosening of the soil and pruning of underground plant organs in the cutouts, performed by cultivators with plane-cutting specially placed paws.

Combined aggregate processing is a method of a complex of techniques that facilitates the combination of several technological operations of tillage (crumbling, loosening, levelling, compaction). It is carried out by tillage units such as CD-720 M, OP-8.5, grain seeder SKP-2.1, rowed seeder SKP-2.1 M, drill no-till, AKSH and RVC, etc. soil, plant residues, fertilizers by rotating milling workers—careful crumbling, loosening, mixing by milling organs. Average tillage is the effect of tillage machines on the soil in a certain way to a depth of 16–25 cm. Ploughing is a method of tillage with a ploughing, which ensures crumbling, loosening and wrapping of the treated soil layer by at least 135°. The main purpose of dump ploughing is to restore high fertility in the entire arable layer. When ploughing with ploughs with plough shares, the latter dump the top layer of soil to the bottom of the furrow, and the main body of the plough lifts the lower well-crumbling part of the arable layer and covers the top layer with it. With such ploughing, deep sealing and decomposition of all plant residues, pests and germs of diseases are carried out, and weeds are deeply pruned.

Ploughing is a method of tillage with a ploughing, which ensures crumbling, loosening and wrapping of the treated soil layer by at least 135°. The main purpose of dump ploughing is to restore high fertility in the entire arable layer. Ploughing with a ploughing with a 180° formation wrap is called a formation turn, with a 135° pa wrap and laying layers at an angle of 45° to the horizon is called a formation sweep. Disk ploughings can also be used for ploughing. The soil is treated in order to increase the capacity of the arable field, to seal up fertilizers, and to loosen the plough sole. Aeration improves, so that plants get enough moisture, and microorganisms are activated. Seedlings appear faster in loose soil. Free loosening of the soil provides crumbling without wrapping with conventional ploughings with removed dumps, chisel ploughings and cultivators. When loosening is not carried out, some part of the stubble remains on the soil surface, pruned weeds and pest larvae, some of the dusty particles located in the upper layer of the soil fall into deeper layers during loosening (FAO, 2011b).

Deep processing is the periodic impact of tillage tools and machines on the soil in a certain way in order to increase the capacity of the treated layer without deep processing is the periodic impact of tillage tools and machines on the soil in a certain way in order to increase the capacity of the treated layer without significantly changing the genetic composition to a depth of 25–35 cm. Ploughing with ploughing of the underlying soil layer.—provides wrapping, crumbling, loosening, bringing to the surface of a part of the podzolic horizon, pruning and embedding in the soil of aboveground organs of weeds, post-harvest residues of cultivated plants, fertilizers, weed seeds, rudiments of diseases and pests of cultivated plants. This technique is used to increase the capacity of the arable layer of sod-podzolic soils, newly developed peat soils (Faulkner, 1943).

Chisel processing is a type of processing, loosening, crumbling of arable and sub-arable horizons without turnover of the formation. Chisel loosens the soil, tearing it from the monolith, but does not compact the under-arable layers and does not form a "ploughing sole". By cutting through the cracks, it promotes better absorption of water by the soil, deeper penetration. According to the depth of soil loosening, chisel tools are divided into cultivators, ping is a type of processing, loosening, crumbling of arable and sub-arable horizons without turnover of the formation. Chisel loosens the soil, tears the monolith but does not compact the under-arable layers, does not form a "ploughing sole", by cutting through the cracks.

By reducing the number of passes over the surface of arable land, that is, by combining the number of operations, minimizing soil cover, protecting, as well as the soil surface with mulch, which is the basis for soil vital activity, nature conservation agriculture creates an ecological base in the farming system, thanks to which it is necessary to create sustainable agricultural production, as well as a land management strategy. Many scientists in America, Asia, as well as in Europe, show that the use of resource-saving technology strategies can mobilize greater potential of crop production, livestock and land resources in the agricultural production system, at the same time contribute to the provision of ecosystem services such as the availability of clean water, control of runoff and soil erosion, pollination services, as well as nutrients, and the carbon cycle (Fig. 9.2) (Farooq & Nawaz, 2014). The effects of conservation agriculture practices on the whole living system in the soil can be explained as shown in Fig. 9.2: (a) Prevention of different stages of insect pests to direct exposure to sunlight or high temperature and their natural enemies, (b) weed as an alternate host for insect pests in crop fields, (c) reduced tillage supporting life cycle of soil-dwelling pests, (d) straw mulch, (e) living mulch, (f) mulching impairing with insect's ability to locate their host plant, (g) organic amendments making soil favourable for pests, (h) organic amendments supporting insect migration and pupation inside soil, (i) intercropping influence on pests density, (j) cover crop influencing pests density, (k) intercropping, a barrier to insect- pests to locate host plant.

As can be considered from the above, soil-saving tillage systems provide greater productivity of production factors and profit, as well as efficiency and have increased resistance to biotic and abiotic stresses compared to traditional tillage systems. Many countries of the world consider it favourable for climate change both in terms of adaptation and mitigation. Environmental protection treatment (EPT) of the soil includes a wide range of methods, the purpose of which is to leave a certain amount of plant residues on the soil surface after harvesting, which increases the penetration of water into the soil and reduces erosion. In modern literature, about 14 names of abbreviated tillage are used: zero tillage, chemical ploughing, treatment with a seed deposit, chemical steam, without tillage, direct sowing, disk drilling, age of the seeder, sowing of turf, minimal tillage or abbreviated tillage, strip tillage, ridge treatment. In general, after harvesting, both the seeds and the plant stem remain on the soil surface. This means that over time all these are used as natural fertilizers and also enrich the soil. Surface coating with residues characterizes the lower limit of the EPT classification, but other conservation goals may include saving time, fuel, soil, and substances. Figure 9.3 shows (a) conventional inversion mould board plough, (b) par plough, (c) deep ripper combination tool, (d) deep field cultivator combination

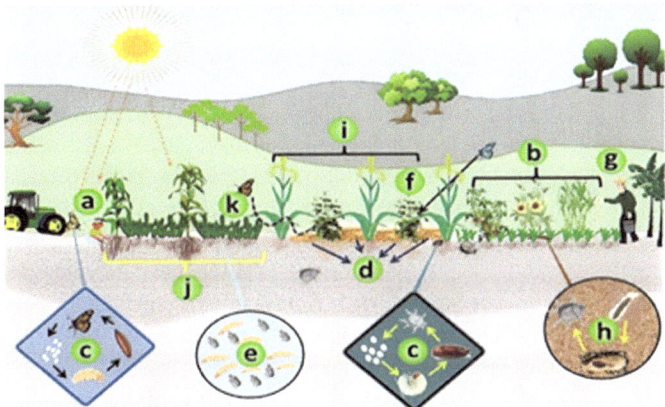

Fig. 9.2 Overall pictorial view of impact of conservation agriculture practices on insect pests and their natural enemies (Farooq & Nawaz, 2014)

Fig. 9.3 Collection of several different types of "conservation tillage" tools and (planters illustrating a wide range of tillage depths and degrees of residue (Farooq et al., 2011)

tool, (e) field cultivator tool rigid tine and (f) deep field cultivator spring tine (NRCS) (Farooq et al., 2011).

Nature conservation tillage (NCT) is a lower intensity of tillage compared to a dump ploughing, which includes a wide range of soil disturbances and the sealing of plant residues. Non-fallow or direct sowing is often included in a broad class of NCT, but perhaps it should be considered as a separate class for NCT with minimal quantitative soil disturbance. When we consider the soil as a living system, it is necessary to take into account the minimum impact on the soil and the introduction of organic substances in order to achieve the goals of improving the condition of the soil and its protection.

Chisel ploughings have passive rigid or spring teeth and usually work to a depth of 100 to 150 mm, stirring the soil and leaving some plant residues on the surface. When using combined tools for deep loosening, subsurface cuttings are used, which penetrate to a depth of about 380 mm, and heavy discs for cutting off residues and levelling the soil surface.

A large number of soil surface disturbances affect fuel consumption, because with less impact on the soil, less fuel is consumed. The main purpose of protecting the soil from residual precipitation in the amount of 30–45% is to preserve soil and water; however, soil conditions can be very diverse. (Fig. 9.4) (Faulkner, 1943).

One of the mandatory spring agricultural practices is cultivation. This is a continuous or row-to-row ploughing without turning over the lower moistened layer, levelling the ridged surface after ploughing. It is necessary to create a dense seedbed for planting so that the seeder coulters do not seal the seeds deeply. When planting seeds in an unprepared field, you cannot count on friendly shoots for three reasons: (i) When sown at different depths, they germinate unevenly, (ii) later shoots are weaker and more susceptible to diseases and (iii) the ridged surface evaporates more moisture. In the heat, the top layer of the soil dries up quickly and deprives the roots of moisture. The aisles clog weeds and take away nutrients from crops.

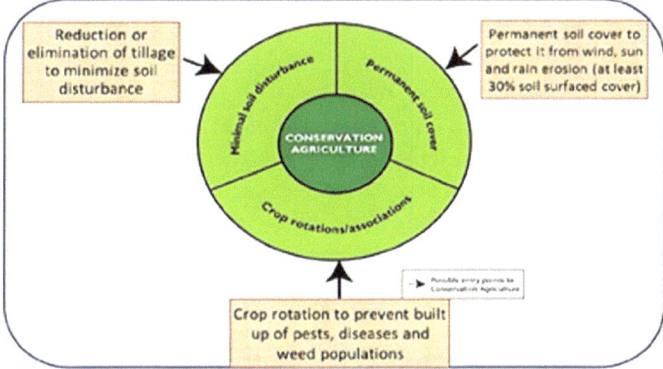

Fig. 9.4 Three principles of conservation agriculture (Faulkner, 1943)

All these disadvantages are solved by mechanical processing equipment. The impact of paws and knives on the arable layer creates favourable conditions for the growth of crops and increasing yields (Farooq et al., 2011).

Considering the advantages of spring pre-sowing preparation, the undoubted advantage of the agricultural approach is the destruction of the soil crust by 0.6 cm and clumped small lumps with voids forming chains of capillaries. In summer, the water rises quickly and evaporates through them. The continuous process leads to the fact that about 100 L of moisture are consumed from 1m^2 on a hot day. Half as much evaporates from the levelled, finely lumpy earth. In addition, dense lumps prevent the flow of air and nutrients to the roots. This significantly impairs the development of plants and often leads to death. During surface loosening, fertilizers are embedded in the arable layer. This has a positive effect on the mineral nutrition of the soil and the foci of the spread of microorganisms. Actinomycetes, aerobes, fungi and microbial colonies improve humidity, aeration, soil biological and physico-chemical processes.

On the other hand, the pre-sowing treatment method is performed twice: the first time in advance at a depth of up to 12 cm; the second—before sowing at 3–6 cm. The cultivator cuts the roots of weeds, but does not bring the lower and wet layers of the soil to the surface. First, a tractor or a monoblock loosens the field horizontally in the direction of ploughing and diagonally; again—across the previous one. The main ways of movement during cultivation are shuttle and diagonal-angular. In the first case, the tractor passes the line at a distance equal to 1.5 of the width of the cutter. The remaining plots are cultivated together with the turning strips. The second method involves processing according to the scheme, followed by beating off the turning lines from all sides. For a unit with a wide grip and on short runs with no possibility of leaving the field, a shuttle method with a loop-free turn is chosen.

9.3.4 An Economic Rationale for Promoting Conservation Agriculture

The agricultural model based on mechanical tillage, open soils and ongoing mono-culture farming has such serious negative consequences for the natural resource base of agriculture that the future production potential of agriculture is under threat. According to experts, this farming system is the main cause of the loss of biodiversity; it accelerates soil degradation by increasing the mineralization of organic substances and the rate of erosion. Healthy soils are the key to building sustainable crop production systems that are resilient to the effects of climate change. This contains a diverse community of organisms that help fight plant diseases, insect pests and weed populations, recycle soil nutrients and improve soil structure, having a positive effect on moisture retention, nutrient retention and intake, as well as on organic carbon levels. To solve these problems, one of the adaptation options is resource-saving agriculture, which is usually determined by three management principles: minimal mechanical disturbance of the soil, permanent organic soil coverage and diversification of crop types due to various sequences and crop associations.

Conservation agriculture is based on three interrelated principles adapted to local conditions and needs: Minimal mechanical disturbance of the soil (i.e. no/no tillage) due to direct application of seeds and/or fertilizers. This reduces soil erosion and preserves soil organic matter. Replacing one cultivation, there is minimal loosening of the soil, due to which the physical and mechanical structure is preserved, as well as when 30 percent of the crop remains on the field surface after harvesting. Maintaining a protective layer of vegetation on the soil surface suppresses weeds, protects the soil from the effects of extreme weather conditions, helps maintain soil moisture and prevents soil compaction. Diversification of agricultural plant species through different crop sequences and associations involving at least three different types of crops can control weeds, as well as the level of soil fertility. A well-thought-out crop rotation creates the basis for a good soil structure, promotes a variety of soil flora and fauna, which provides a nutrient cycle and improves plant nutrition, as well as helps protect them from pests and diseases. Environmental agriculture is quite widely developed and has proven its effectiveness. For example, in Mexico, legumes in crop rotation with corn introduce organic substances and nitrogen into the soil, which help increase corn yields by 25%. Zero tillage contributes to an increase in wheat yields in the range from 6 to 10%, as it allows for timely sowing, leads to an improvement in the condition of crops and provides significant savings in the operation of agricultural machinery, as well as time and fuel. In the west of the Indo-Gangetic Plain, the introduction of zero-tillage in wheat production reduced farmers' costs per hectare by 20% and increased net income by 28%. Conservation agriculture is 20–50% less labour-intensive and thus contributes to reducing greenhouse gas emissions by reducing energy consumption and increasing the efficiency of the use of nutrients. At the same time, it stabilizes and protects the soil from destruction and carbon release into the atmosphere. The Food and Agriculture Organization of the United Nations (Paarlberg, 2001) promotes the introduction of resource-saving agriculture and environmental protection measures. FAO's support to Member States includes several areas:

1. Farming methods and sustainable mechanization. Develop, formulate and plan national strategies and policies that encourage farmers to adopt resource-efficient farming methods and invest in sustainable agricultural mechanization. This allows farmers, especially small ones, to move from inefficient methods of agro-technical management and manual labour to appropriate levels of mechanization that provide higher returns.
2. Training of farmers and service providers, development and dissemination of educational materials and guides to raise awareness, inclusion in the curricula of agricultural universities and government programs.
3. Implementation of location-specific methods and identification of suitable crops to improve production systems that are resilient to the effects of climate change, as well as identification of existing or potential markets for resources and/or products. Increase agricultural production through the introduction of lean farming practices in support of national priorities related to food security and nutrition.

The functioning of resource-efficient agriculture can also be improved through the use of advanced agricultural practices, some of which include planting stress-resistant crop varieties and adequate nutrient provision (Perszewski, 2005). Resource-saving agriculture changes the properties and processes of the soil. These changes, in turn, can affect the provision of ecosystem services, including climate regulation through carbon sequestration and greenhouse gas emissions, as well as regulation and provision of water through the physical, chemical and biological properties of the soil. There is sufficient evidence that the content of organic matter in the upper soil layer increases with resource-saving agriculture, and at the same time, water quality increases due to soil properties and processes that reduce erosion and surface runoff. According to experts, the impact on other ecosystem services is less obvious. Only about half of the more than 100 studies comparing carbon sequestration in soil with zero and traditional tillage showed an increase in sequestration with zero tillage; this is despite constant claims that conservation agriculture captures soil carbon. The same can be said about other ecosystem services. Some studies report higher emissions of greenhouse gases (nitrous oxide and methane) from resource-saving agriculture compared to traditional agriculture, while others find lower emissions, Moisture retention in the soil may be higher with resource-saving (environmental protection) agriculture, which leads to higher and more stable yields in dry seasons, but the amount of residues and the level of soil organic matter necessary to achieve a higher moisture content in the soil are unknown.

In Fig. 9.5, households make technology choices and decisions about the use of their soil resources under the constraints imposed by their socio-economic attributes and on-farm resources, as well as higher-level factors at the local to global scales (Putte et al., 2010). For example, lacking adequate tenure and access to credit, the farmer cannot invest in CA if this requires a large capital outlay. Information about new technologies and financial conditions is a precursor to changes in farm practices and acquiring it does not usually involve large financial outlays. Government

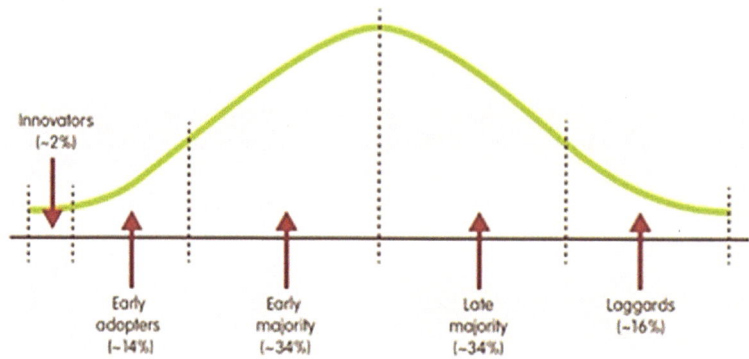

Fig. 9.5 Bell-shaped curve showing categories of individual innovativeness and percentages within each category (Thierfelder & Wall, 2009)

credit and extension policies play an important role here. In contrast to the more direct working of agriculture sector policies and financial incentives, some social and institutional factors have a more indirect influence (Reicosky & Saxton, 2007). Nonetheless, all these factors affect the net returns, risks and other pecuniary elements that drive the decision-making process.

The working of the feedback mechanisms (Fig. 9.5) closes the loop, and there is the potential for either a self-reinforcing series of improvements in soil productivity, or spiraling degradation.

9.4 Conclusion

The main advantage of CA is that it does not use dangerous chemicals from conventional agriculture, which contribute to the occurrence of many serious diseases, including cancer; a decrease in fertility, immune resistance of the human body; depressive and aggressive conditions, as well as many other negative effects.

Another advantage of CA is that the resulting fruits, vegetables and animal products contain much more nutrients needed by the human body and ecological agriculture has many other advantages over traditional agriculture. Among the most important, the following can be noted: (i) New opportunities for economic development are emerging; more expensive products, traditional occupations, agroecological tourism, a greater variety of products and the creation of new market niches, (ii) long-term soil fertility is maintained without the cost of mineral fertilizers, (iii) disease and pest control is carried out without destroying nature, (iv) the purity and safety of water used for drinking and irrigation are maintained, (v) there are more diverse, delicious and nutritious foods, (vi) The health problems of farmers and consumers associated with the use of chemicals, antibiotics, hormones, etc. are reduced, while reducing the associated costs, (vii) losses in natural disasters are reduced, as ecological plants have a more powerful root system, and the soil is less susceptible to drought, having a better structure, (viii) more variety of products for consumption by your own family, which contributes to the health of the farmer's family and (ix) increasing the profitability of low-investment technologies.

References

Çetin, Ö., Üzen, N., & Koca, Y.K. (2018). The importance of amelioration on soil and water resources in the rural areas of Turkey. *International GEA (Geo Eco Eco Agro) Conference*, 1–3 November, 2018, Podgorica, Montenegro. Book Proceedings, pp. 146–152.

Cynthia, L. F., Joyce, S. F., Gerald, R. G., Sarah, B. G., Nancy, L. H., & Sherry, C. W. (1983). Expression of bacterial genes in plant cells. *Proceedings of the National Academy of Sciences of the United States of America, USA, 80*, 4803–4807.

Derpsch, R. (2003). Conservation tillage, no-tillage and related technologies. In G. T. Luis, B. José, M. V. Armando, & H. C. Antonio (Eds.), *Conservation agriculture: Environment, farmers experiences, innovations, socio-economy, policy* (pp. 181–190). Springer.

Dill, G. M. (2005). Glyphosate-resistant crops: History, status and future. *Pest Management Science, 61*, 219–224.

Dumanski, J., Peiretti, R., Benetis, J., McGarry, D., & Pieri, C. (2006). The paradigm of conservation tillage. In: *Proceedings of World Association of Soil And Water Conservation*, 58–64.

FAO. (2008). Investing in sustainable crop intensification: the case for soil health. Report of the international technical workshop, FAO, Rome, July. *Integrated Crop Management*, 6. FAO, Rome. Retrieved from http://www.fao.org/ag/ca/. Access date May 18, 2014.

FAO. (2011a). CA adoption worldwide, FAO-CA website. Retrieved from http://www.fao.org/ag/ca/6c.html. Access date April 11, 2024.

FAO. (2011b). *Save and grow: A policymaker's guide to the sustainable intensification of smallholder crop production*. FAO.

FAO. (2024). Conservation agriculture. Retrieved from https://www.fao.org/conservation-agriculture/overview/why-we-do-it/en/. Access date May 14, 2024.

Farooq, M., & Nawaz, A. (2014). Weed dynamics and productivity of wheat in conventional and conservation rice-based cropping systems. *Soil and Tillage Research, 141*, 1–9.

Farooq, M., Jabran, K., Cheema, Z. A., Wahid, A., & Siddique, K. H. M. (2011). The role of allelopathy in agricultural pest management. *Pest Management Science, 67*, 494–506.

Faulkner, E. H. (1943). *Ploughingman's folly*. Michael Joseph.

Fawcetti, R. S. (1987). Overview of paste management for conservation tillage systems. In T. J. Logan, J. M. Davidson, L. Baker, & M. R. Overcash (Eds.), *Effects of conservation tillage on groundwater quality: Nitrates and pesticides* (pp. 19–37). Lewis.

Foresight. (2011). *The future of food and farming*. The Government Office for Science.

Friedrich, T., & Kassam, A.H. (2009). Adoption of conservation agriculture technologies: constraints and opportunities. In: *Proceedings of the IV world congress on conservation agriculture*, ICAR, New Delhi, India, 4–7 Feb 2009.

Friedrich, T., Derpsch, R., & Kassam, A. H. (2012). Global overview of the spread of conservation agriculture. *Field Actions Science Reports, 6*, 1–7.

Garbach, K., Milder, J. C., M Montenegro, M., Karp, D. S., & DeClerck, F. A. J. (2014). Biodiversity and ecosystem services in agroecosystems. *Encyclopedia of Agriculture and Food Systems, 2*, 21–40. https://doi.org/10.1016/B978-0-444-52512-3.00013-9

Komarek, A. M., Thierfelder, C., & Steward, P. R. (2021). Conservation agriculture improves adaptive capacity of cropping systems to climate stress in Malawi. *Agricultural Systems., 190*, 103–117. https://doi.org/10.1016/j.agsy.2021.10311

Miguel, A. F., Peñalva, M., Calegari, A., Derpsch, R., & McDonald, M. J. (2011). Green manure/ cover crops and crop rotation in conservation agriculture on small farms. *Plant Production and Protection Division*, FAO, Rome.

Montgomery, D. R. (2007). Soil erosion and agricultural sustainability. *Proceedings of the National Academy of Sciences of the United States of America, USA, 104*, 13268–13272.

Paarlberg, R. L. (2001). *The politics of precaution: Genetically modified crops in developing countries*. Johns Hopkins University Press.

Perszewski, R. (2005). Ideas leading to no-till's second revolution. Retrieved from http://www.no-tillfarmer.com/pages/Feature-Articles-Ideas-Leading-To-No-Tills-Second-Revolution. Access date June 2, 2024.

Potter, T. L., Truman, C. C., Strickland, T. C., Bosch, D. D., & Webster, T. M. (2008). Herbicide incorporation by irrigation and tillage impact on runoff loss. *Journal of Environmental Quality, 37*, 839–847.

Putte, A. V., Govers, G., Diels, J., Gillijns, K., & Demuzere, M. (2010). Assessing the effect of soil tillage on crop growth: A meta-regression analysis on European crop yields under conservation agriculture. *European Journal of Agronomy, 33*, 231–241.

Reicosky, D. C., & Saxton, K. E. (2007). The benefits of no-tillage. In: Baker C. J., Saxton K. E., Ritchie, W.R., Chamen, W. C., Reicosky, D. C,, Ribeiro, M. F., Justice, S. E, Hobbs, P. R (eds.) No-tillage seeding in conservation agriculture. 2nd edn. CABI, Wallingford, 11–20.

Serraj, R., & Siddique, K. H. M. (2012). Conservation agriculture in dry areas. *Field Crops Research, 132*, 1–6.

Thierfelder, C., & Wall, P. C. (2009). Effects of conservation agriculture techniques on infiltration and soil water content in Zambia and Zimbabwe. *Soil Tillage Research, 105*, 217–227.